IMAGES
of America

FIREFIGHTING IN FORSYTH COUNTY

On the Cover: The original Forsyth County fire jeep is shown here with county commissioners William Simpson, Wally Dunham, Dr. D.C. Speas, and Carl Ogburn. (Courtesy of Forsyth County Public Library.)

IMAGES
of America

FIREFIGHTING IN FORSYTH COUNTY

Joshua Wright

ARCADIA
PUBLISHING

ISBN 978-1-4671-0712-9

Published by Arcadia Publishing
Charleston, South Carolina

Printed in the United States of America

Library of Congress Control Number: 2021947028

For all general information, please contact Arcadia Publishing:
Telephone 843-853-2070
Fax 843-853-0044
E-mail sales@arcadiapublishing.com
For customer service and orders:
Toll-Free 1-888-313-2665

Visit us on the Internet at www.arcadiapublishing.com

This book is dedicated to all the firefighters and emergency responders in Forsyth County, North Carolina, past, present, and future.
May you always be courageous and find peace at the end of the day.

Contents

Acknowledgments

Thank you to all the Forsyth County volunteer fire departments featured in this book. Thank you also to the Winston-Salem Fire Department and Kernersville Fire Department. Special thanks to fire chiefs Tim Whicker, City View Fire Department (CVFD); Steve Williams, Lewisville Fire Department (LFD); Barry McLean, Forbush Volunteer Fire Department (FVFD); Daren Zigler and Ross Flynt, Walkertown Fire Department (WFD); Norman Walker, Mineral Springs Fire Department (MSFD); Brandon Watson, Salem Chapel Fire Department (SCFD); Kelly Baker (Horneytown Fire Department); Gary Styers and Chris Vogler, Forsyth County Fire Department (FCFD); and Capt. Bill Buwalda, Clemmons Fire Department (CFD). Thank you to Capt. David "Pyle" Koehler, Forsyth County Fire Department (FCFD); and Jason Gore, Winston-Salem Fire Department (WSFD), for your rescue history, insights, and lessons. Thanks also go to Chief Carr Boyd, Forsyth County EMS (FCEMS); Dwayne Livengood, Forsyth County Fire Department (FCFD); Aaron Greene; Ben Perkins, City View Fire Department (CVFD); Molly Rawls and the Forsyth County Public Library (FCPL); Old Salem Museums and Gardens; the Kernersville Museum; the Forsyth County Fire Historical Society (FCFHS); Forsyth County Emergency Services; and my supportive family.

Introduction

Whether you believe the Greek myth of Prometheus stealing fire from the gods and giving it to humans, Native American cultures' belief that a wolf or coyote brought fire to humans after stealing it from an evil creature, or simply that naturally curious humans accidentally discovered it, fire has long been an integral part of our lives. Even the scientific world cannot decide on when exactly man first gazed at the natural wonder. It is impossible to know its full history. What we do know is that fire is one of the greatest and most important discoveries made by humans, allowing us to build civilizations and discover new worlds. But at some point, it became obvious that fire needed to be controlled.

This is where the firefighter becomes important. As humans have developed a society, we have sought ways to protect our people and property. The first organized fire brigade were the *vigiles* (watchmen) of Rome. Instituted by Caesar Augustus, the brigade consisted of ex-slaves who were responsible for combating fires using buckets and local water sources. Later Europeans had a similar model. King Louis IX of France issued a decree that allowed citizens to establish their own night watches. This model was instituted across Europe. The method of firefighting did not change until after the Great Fire of London, with pumps and hoses becoming tools for fire protection in the 17th century. The migration of humans all around the world has driven fire protection to advance far beyond anything Caesar or Louis IX could have imagined.

The New World was the frontier for such innovation, where free-thinking ideology allowed for a new level of fire protection. The first recorded structure fire in the Jamestown colony destroyed the building and the majority of the possessions in it in the winter of 1608. The early settlers knew their undertaking had to be protected. America continued to prevail, and towns were formed and fire brigades were established. The need for fire protection was ever-evolving, sometimes outweighing what was available.

The Forsyth County we now know was originally settled by members of the Moravian Church in 1753. Since its inception, Forsyth County has been the birthplace of innovation. The first Moravian settlers drew from their familiarity with German waterworks systems to build one of the first public waterworks systems in the country. The pipes were cut from local trees in the 16th century. This innovative mindset later led to the construction of the Reynolds Building skyscraper, which was the inspiration for the larger Empire State Building in the 20th century. Innovation and dedication to a cause to better human existence have always been virtues of the citizens of Forsyth County.

Forsyth County is located in North Carolina's Piedmont region. It includes Winston-Salem, Kernersville, Clemmons, Lewisville, Tobaccoville, Walkertown, Rural Hall, and many other unincorporated communities and crossroads.

In fire protection, Forsyth County has a history and experience second to none. From the first fire company and bucket brigade organized in 1772 in Salem to the new Kernersville Station 42 with a state-of-the-art training ground, firefighters in Forsyth County have always adapted and overcome when faced with the ever-changing nature of their profession. The fire service in Forsyth County has always been at the forefront of technology, diversity, and revolution. Since the very first fire recorded in Salem, the citizens have stepped up when called to protect lives and property.

In 1753, the Moravians were getting settled in the town they called Salem. This community was very well organized. They established laws to govern and protect their community. The first home in Salem was built in 1766, and shortly after, the community established the first fire department. In 1782, Salem established a Fire Inspector's Committee to assure that chimneys were swept and fires were watched. Fire protection was a top priority. The town purchased the first fire pump in the state in 1785 after a tavern burned down the year before. Old Salem's Market and Fire House is the oldest enginehouse in North Carolina, built in 1803. In 1849, the town of Winston was established north of Salem. Winston, like Salem, formed a volunteer fire department. The towns of Salem and Winston formed into one city on May 13, 1913. Although fire protection for the city of Winston-Salem had made great strides, rural fire protection was still an issue. Many homes, businesses, and churches still had little to no protection. Not until much later did community leaders begin organizing methods for rural fire protection in the county. This book will focus primarily on the rural fire protection of Forsyth County and the men and women who have made it the model it is today. This is their story.

One

Where It All Began

Forsyth County started as a Moravian settlement in 1753. The Moravians called their territory Wachovia after the Austrian estate of Count Nicholas Lewis von Zinzendorf, an early defender of the Moravian Church.

From the beginning, one of the first towns, Salem, was known for its innovation and ingenuity. The townspeople took fire protection seriously. Once the first home was built, the town made plans for handling fires. This led to the Salem Fire Department in 1772. Salem became one of the only towns in the young nation to own a manual fire pump in 1785. In 1849, the town of Winston was formed (named after Joseph Winston, a Revolutionary War hero). This brought more retail and industrial business to Forsyth County, including John Wesley Hanes's Shamrock Knitting Mill (now part of the Hanes company) and the Richard Joshua Reynolds Tobacco Company.

Fire was becoming a rising threat to the two towns. To protect the infrastructure and growing population, the Winston Fire Department was formed in 1876. Salem lost its tavern due to fire in 1782, prompting the building of the Old Salem's Market and Fire House in 1803. The town of Winston suffered the loss of its resort hotel Zinzendorf in 1892. In 1913, the towns of Salem and Winston merged and created the Winston-Salem Fire Department. After the two departments became one, Winston-Salem hired its first full-time paid fire chief, H.E. Nissen, a former member of the Salem Volunteer Company.

In 1912, the city's first motorized fire engine was ordered from American LaFrance. It went into service in the spring of 1913 at Winston-Salem Fire Station No. 2, which also employed the first full-time paid firefighters in the city. Winston-Salem, the largest city in North Carolina in 1920, continued to grow in industry and innovation, and the Winston-Salem Fire Department has stayed true to its legacy. Fire protection and service for its citizens is still of paramount importance.

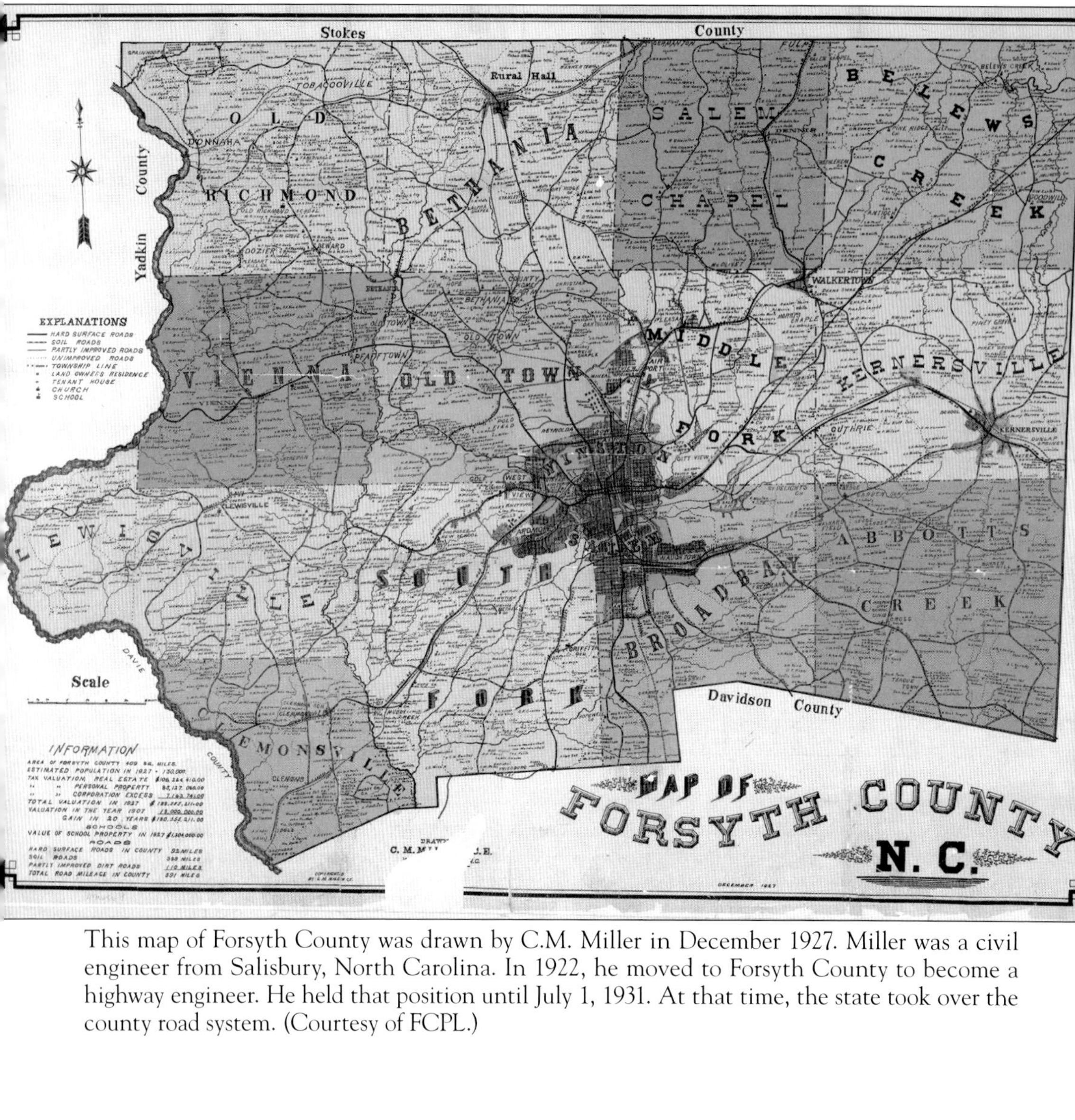

This map of Forsyth County was drawn by C.M. Miller in December 1927. Miller was a civil engineer from Salisbury, North Carolina. In 1922, he moved to Forsyth County to become a highway engineer. He held that position until July 1, 1931. At that time, the state took over the county road system. (Courtesy of FCPL.)

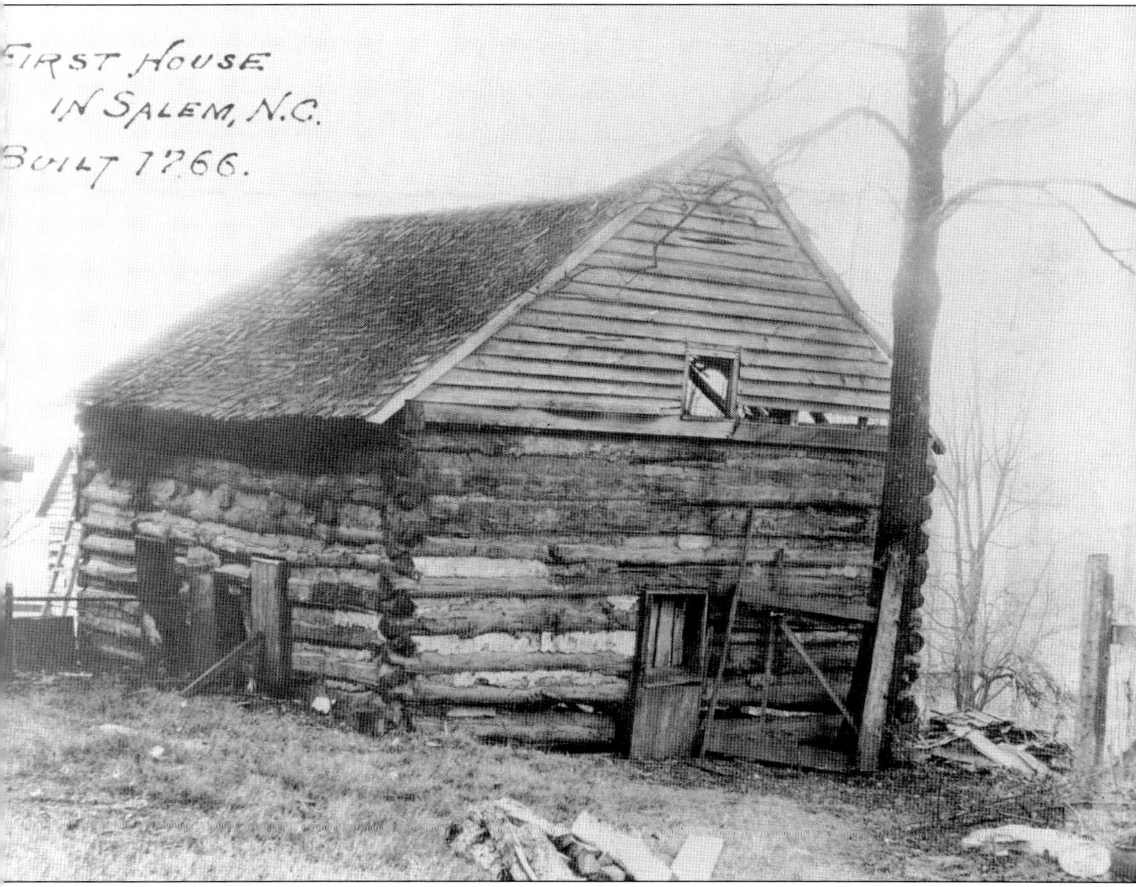

The first home in Salem was known as the Builders' House. It was built in 1766 and fell in 1907. Trees were cut and hewn into logs to build the cabin for the workmen of Salem. (Courtesy of Old Salem Museum and Gardens.)

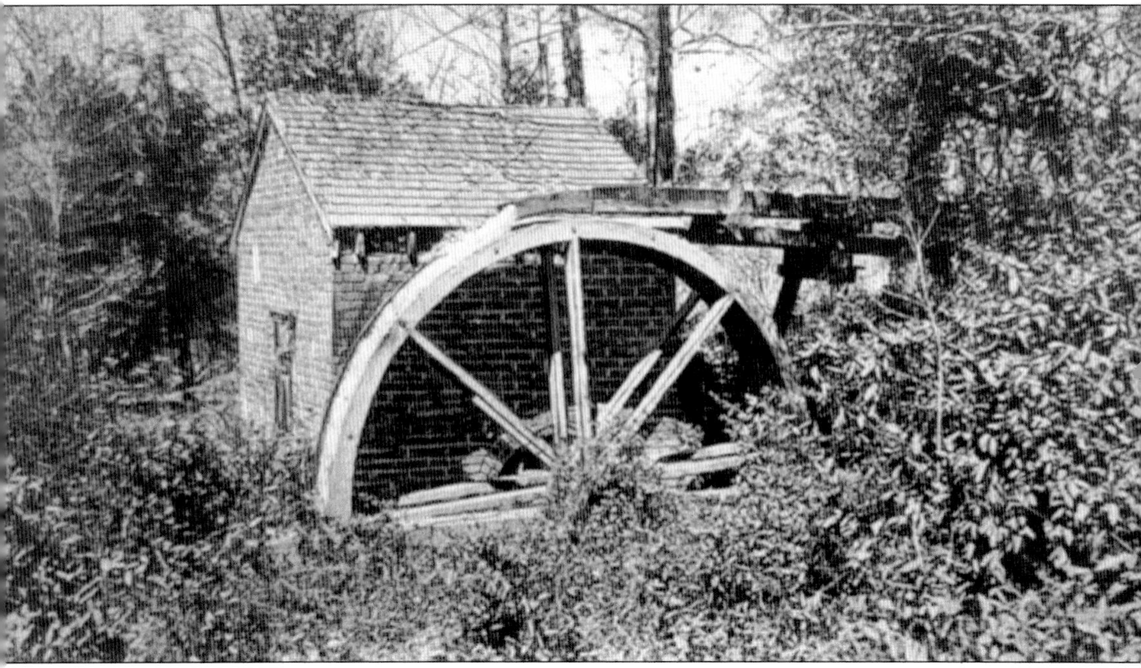

Salem installed the first colonial water systems on December 16, 1778. The Congregation Council employed J. Krause to erect a municipal waterworks system. This system was powered by a waterwheel that fed off Salem Creek. The town was above the flood stages of surrounding streams and below the springs on higher ground. The town could be served with water by gravity. (Courtesy of Old Salem Museum and Gardens.)

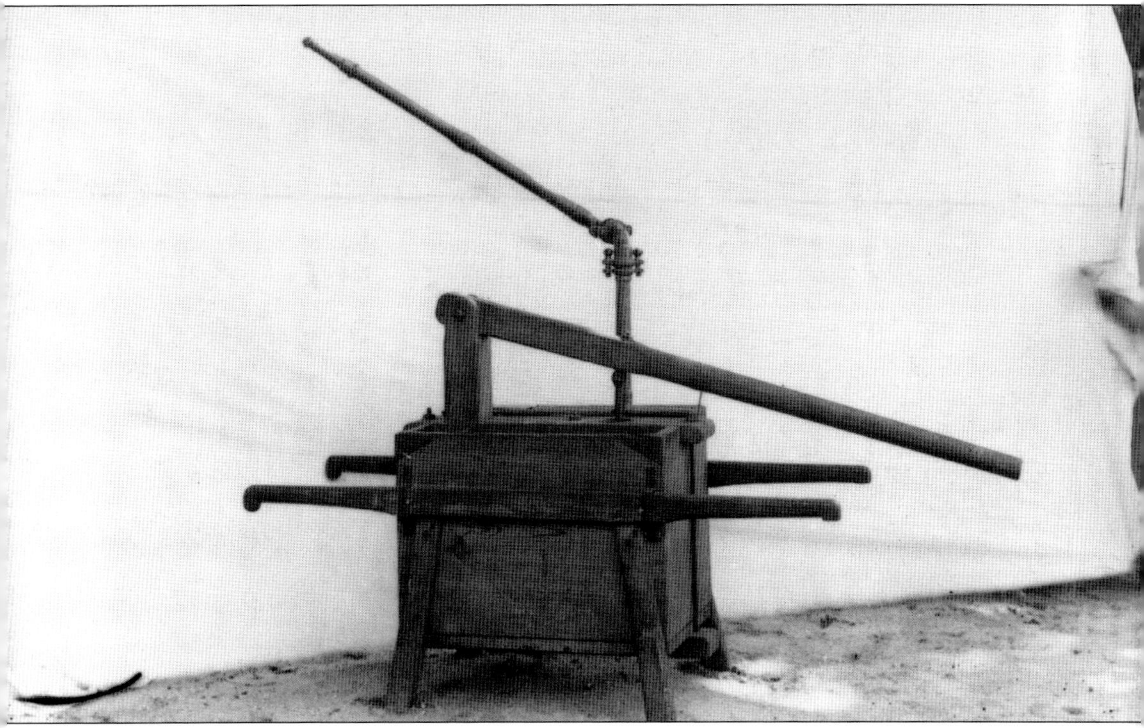

The first fire pump in North Carolina was in Salem. In 1785, Salem became one of the few towns in the entire nation to own a manual fire pump. It had handles on each side so it could be carried to the scene of a fire and loaded with water by a bucket brigade. (Courtesy of Old Salem Museum and Gardens.)

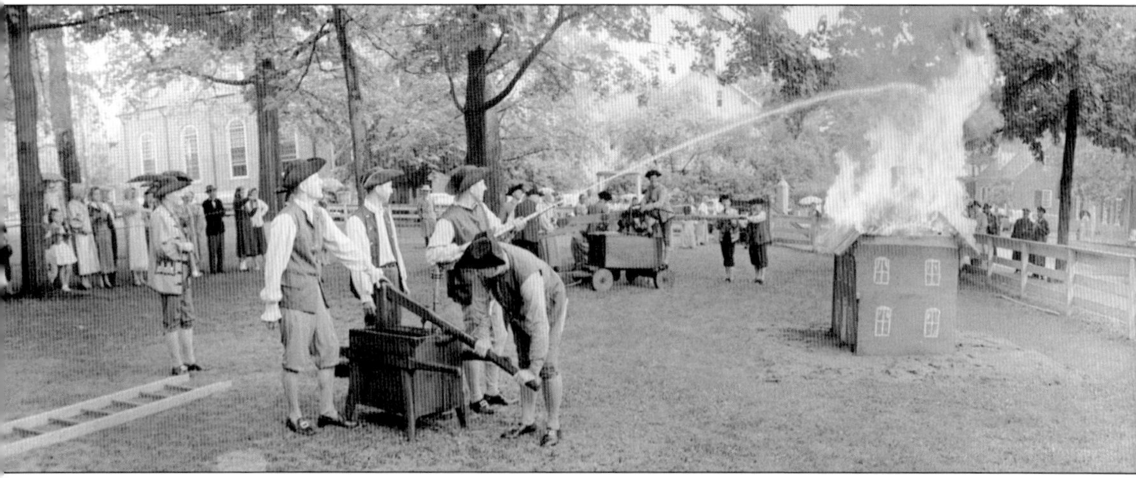

This is the dedication of the Market Fire House in Old Salem in 1957. Using an antique fire engine, the group is reenacting putting out a fire in Salem Square. (Courtesy of FCPL.)

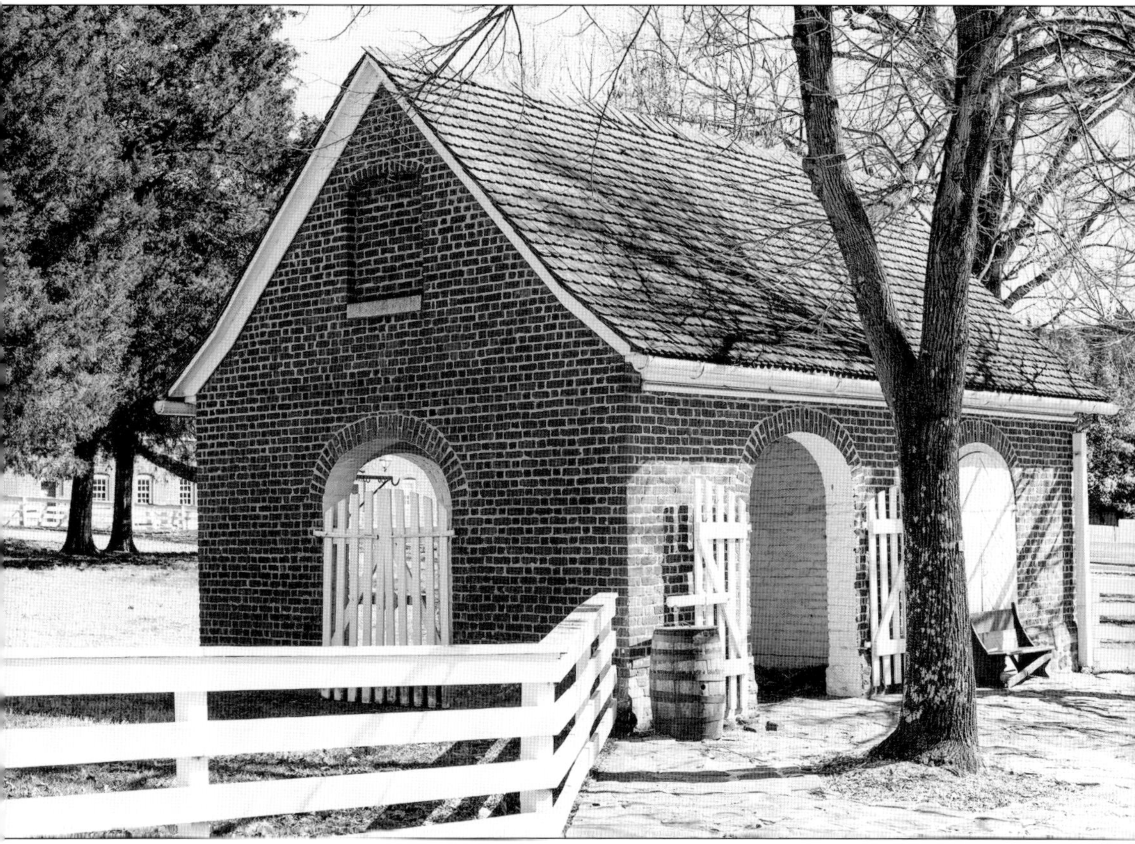

The Market Fire House was built on Salem Square to provide a marketplace for fresh meat. The local doctor had determined that fresh meat was necessary to supplement the salted meat the residents were eating. The other half of the building was used to store the fire equipment that was essential to the town's plan to prevent and manage fires. The Market Fire House is still the oldest standing firehouse in the state of North Carolina. (Courtesy of Sylvia Elizabeth Photography.)

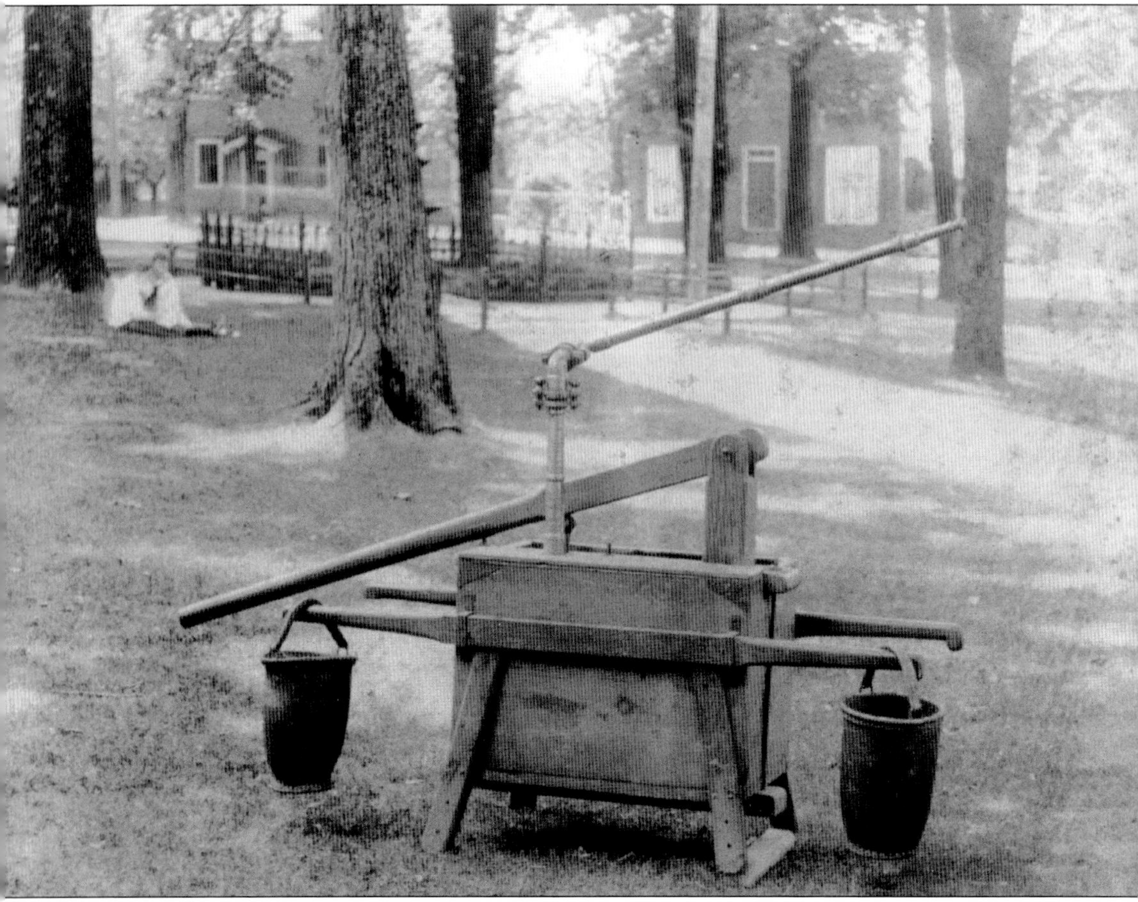

Shown here is a fire engine, ordered nearly a year earlier, that arrived on May 25, 1785. The engine was put into service on June 16, 1785, creating the first fire company in the town of Salem. Even after the engines went into service, Salem's bucket brigades were still used. (Courtesy of FCFHS.)

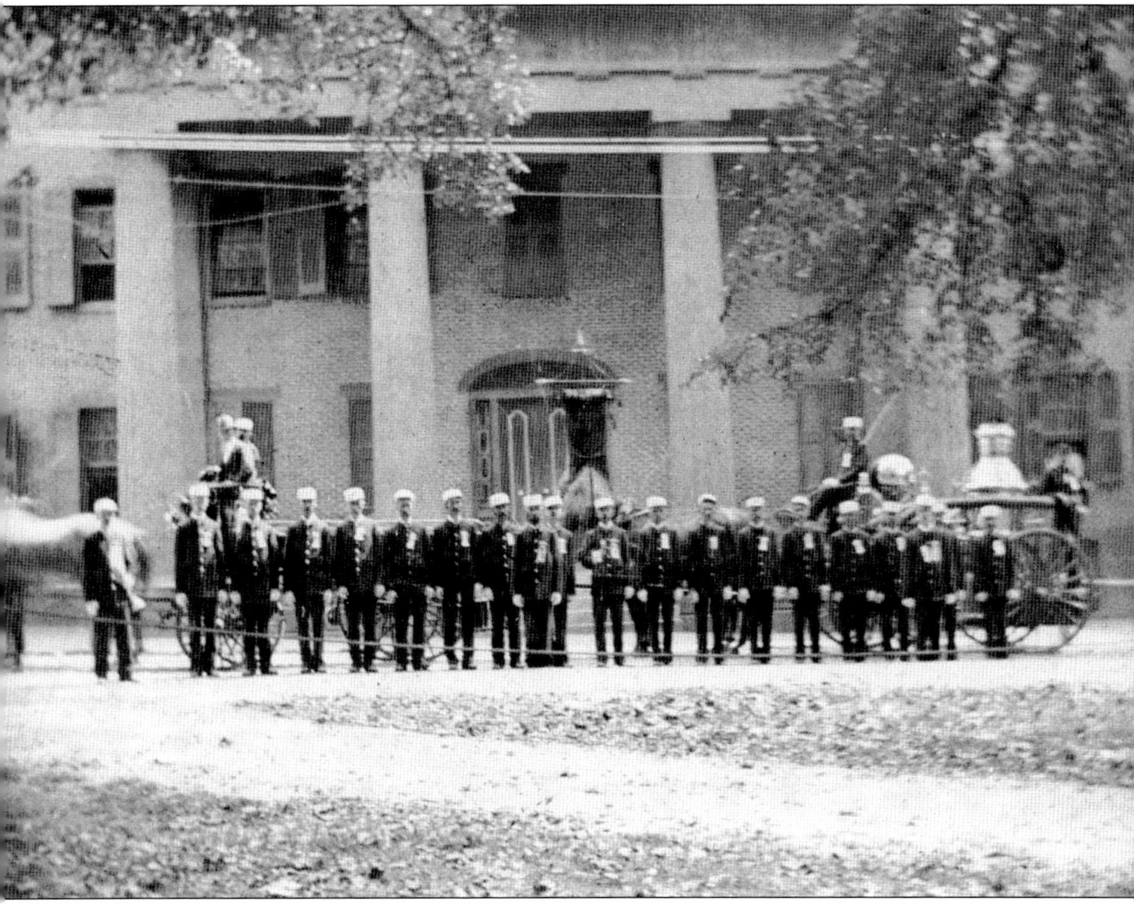

This photograph shows the Salem Fire Department members in front of Main Hall in Salem with their fire truck and fire wagon in 1890. This was one of the first departments to be incorporated in the state. (Courtesy of FCPL.)

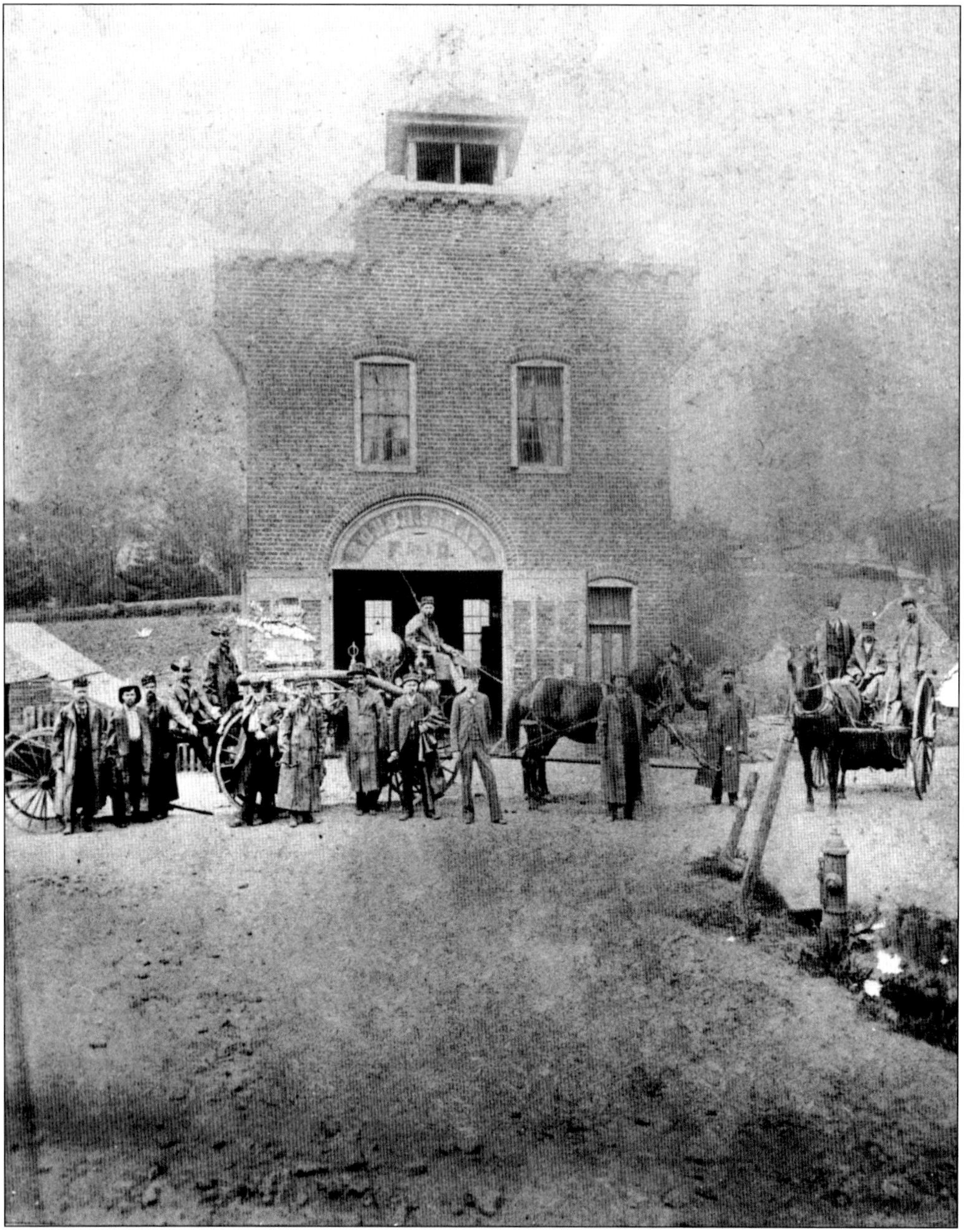

Pictured here is the Button fire engine in front of Salem's Rough and Ready Fire Department at the Eighth Street Fire Station in 1890. The words "Rough & Ready" can be seen over the doorway. The location, as well as the firefighters' quick tactics in firefighting, were unparalleled in the city. Under the leadership of F.C. Meinung, this firehouse set a standard for future generations with their courage and determination. (Courtesy of FCPL.)

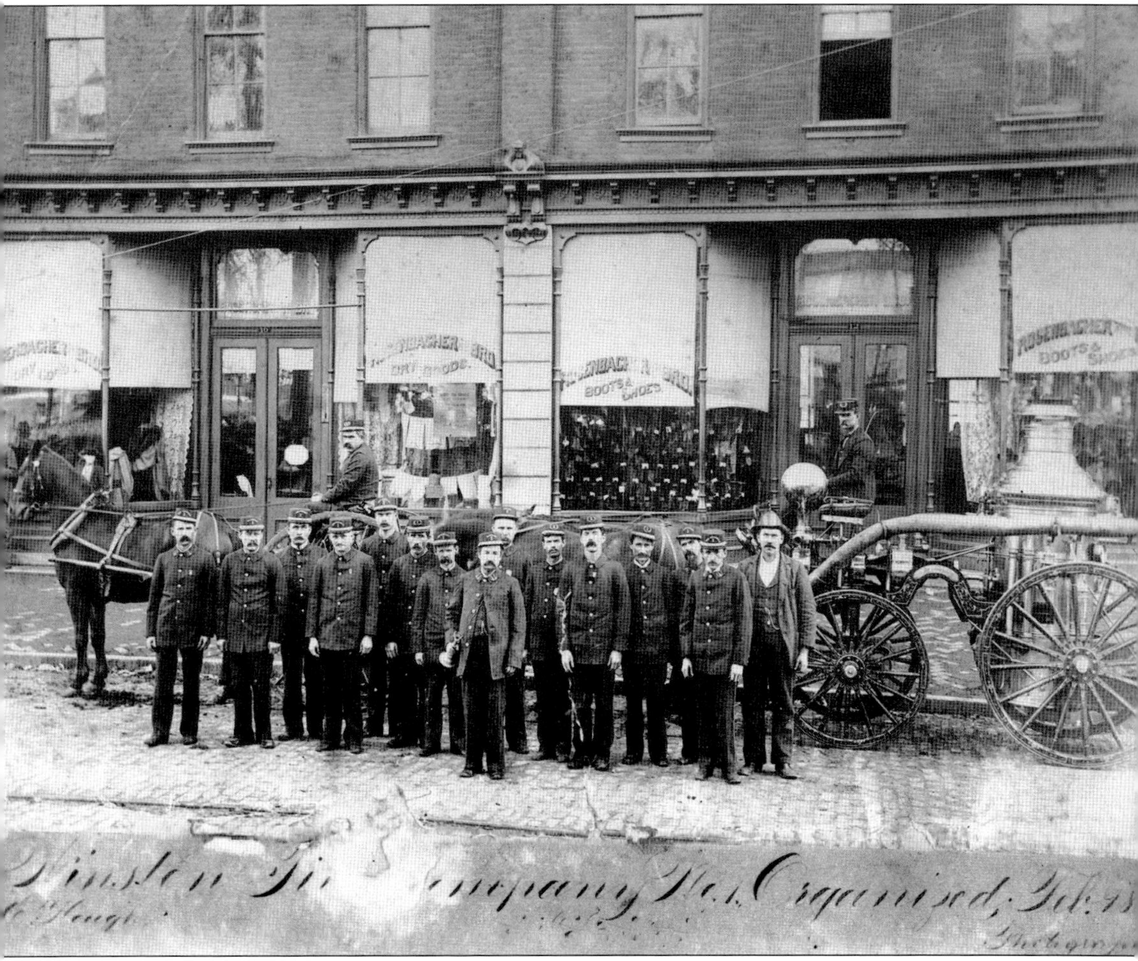

Winston Fire Company No. 1 was organized in 1889. The man in front is Wiley Keith, who organized the company. This photograph was taken in front of the Rosenbacher & Brothers store at 1014 West Third Street. Winston was the industrial capital of the Piedmont area, just north of Salem. (Courtesy of FCPL.)

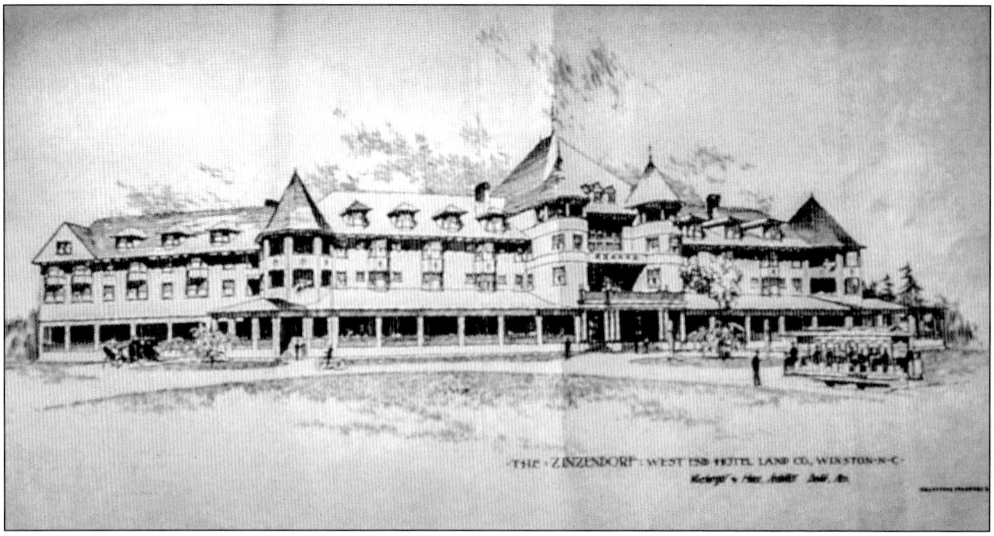

This is the original sketch plan for the Zinzendorf Hotel. The luxury hotel was named after the German-born Count Ludwig Von Zinzendorf, one of the leaders of the Moravian colonists. He had originally supported and laid out plans for the town of Salem. (Courtesy of FCFHS.)

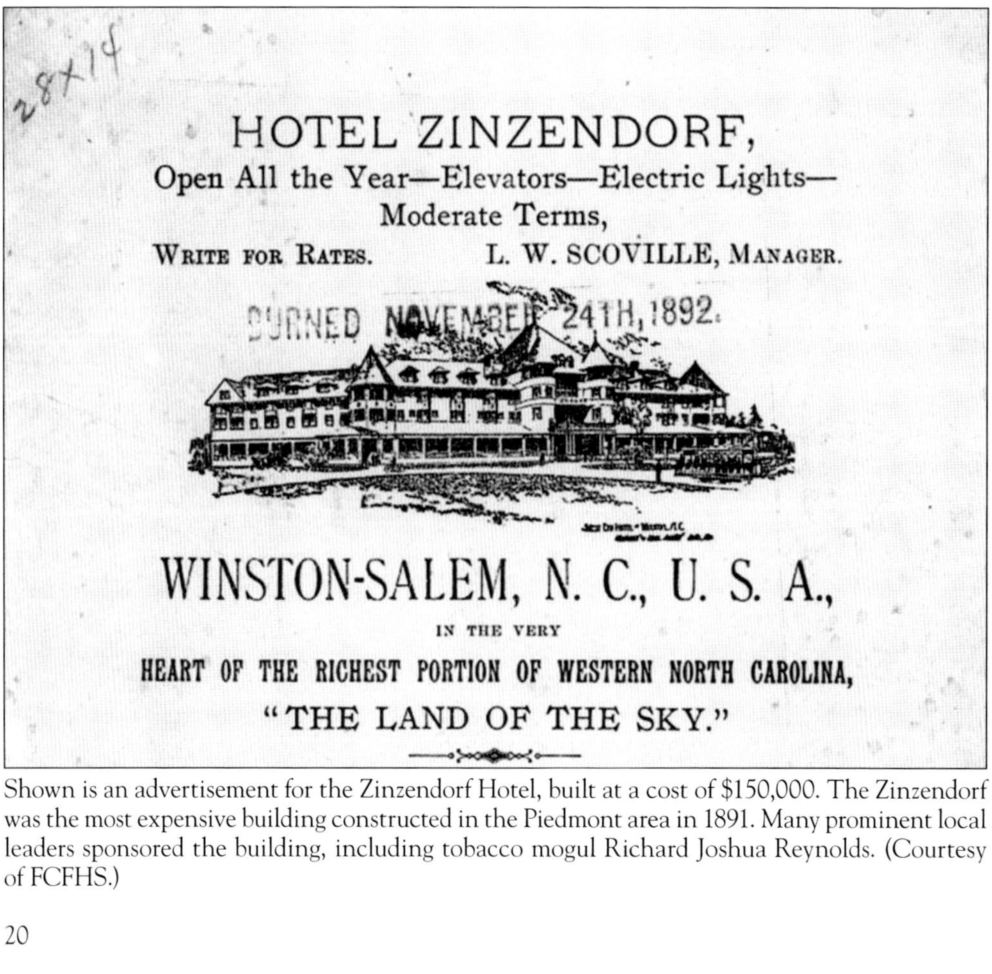

Shown is an advertisement for the Zinzendorf Hotel, built at a cost of $150,000. The Zinzendorf was the most expensive building constructed in the Piedmont area in 1891. Many prominent local leaders sponsored the building, including tobacco mogul Richard Joshua Reynolds. (Courtesy of FCFHS.)

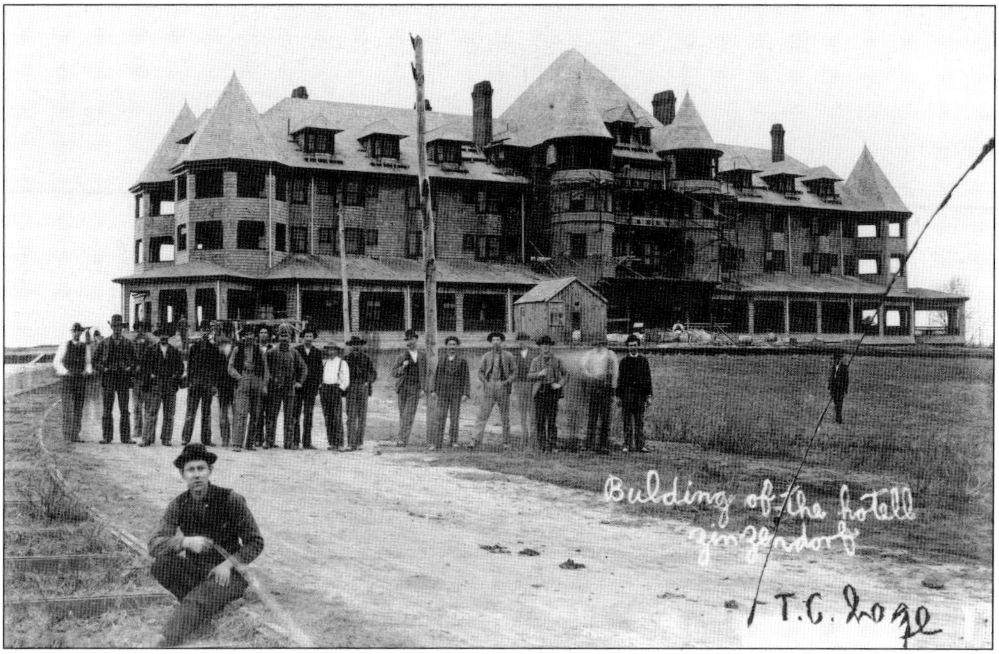

The Zinzendorf Hotel is shown here under construction; it was completed in 1891. Tom Hege, at lower left, is pulling on the string that took the photograph. (Courtesy of FCPL.)

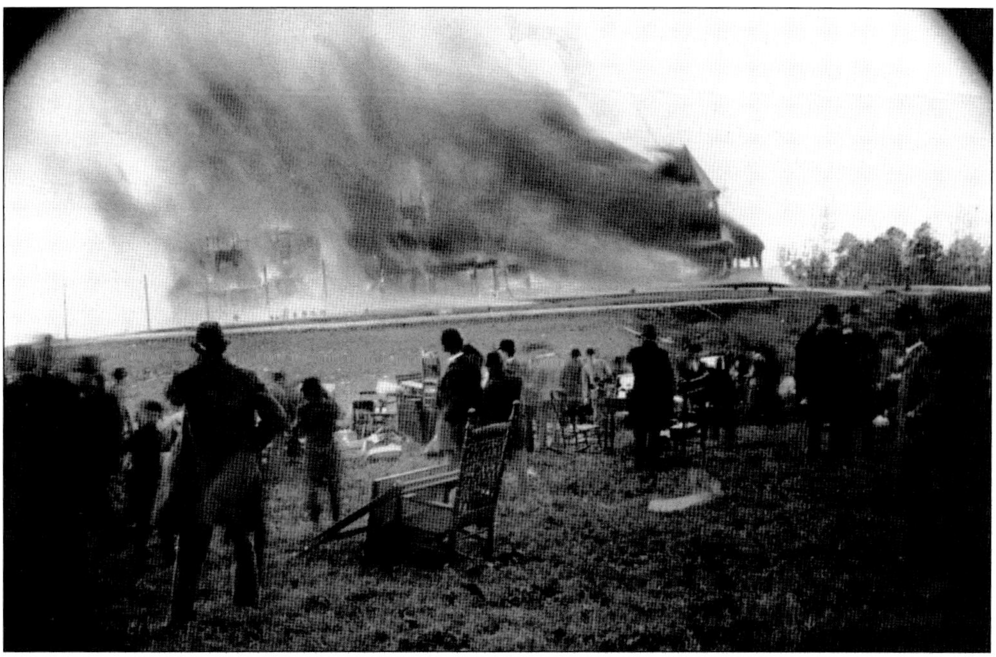

The Zinzendorf Hotel burned on Thanksgiving Day in 1892. Firefighters from Winston and Salem quickly responded. The fire began in the hotel's laundry room. Because of its wooden frame and cedar shingles, the hotel stood no chance against fire. The growing fire could not be contained. Firefighters helped carry out what furnishings and other items could be saved. Everyone escaped injury, but the building was completely destroyed within an hour. (Courtesy of FCPL.)

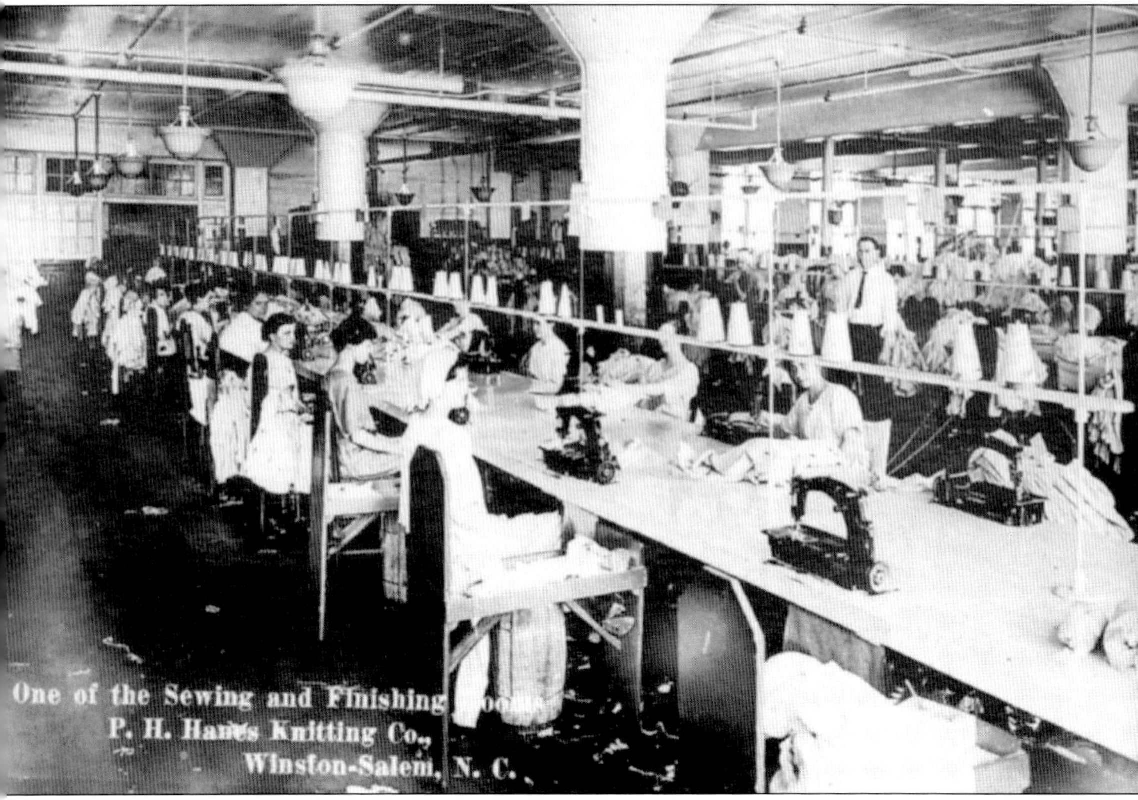

Pleasant H. Hanes founded P.H. Hanes and Co., which manufactured tobacco, from 1872 until 1900, when the company was bought out by competitor Richard Joshua Reynolds Tobacco Company. Hanes then founded the P.H. Hanes Knitting Co. in 1901, which introduced the revolutionary two-piece men's underwear to the public. (Courtesy of FCFHS.)

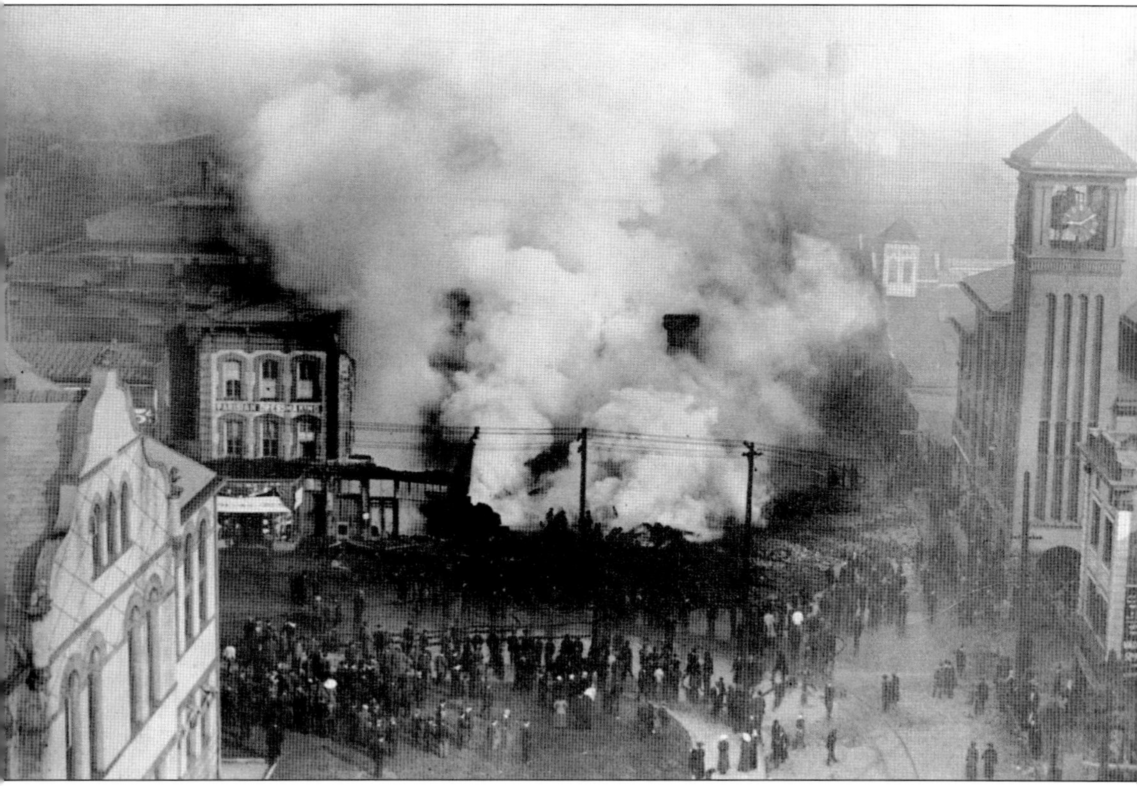

This large fire was at the hardware business of Brown, Rogers & Company on the corner of Fourth and Main Streets in 1913. Also damaged were the Frank C. Brown & Sons Company and the Eagles' Club rooms over the hardware store. (Courtesy of FCPL.)

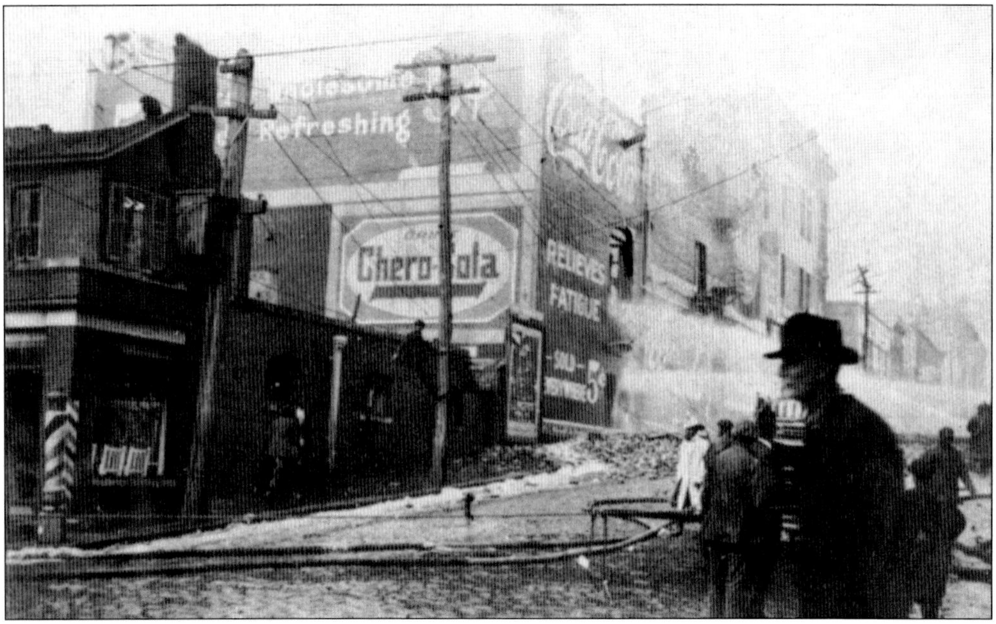

The back and side of the Elks Auditorium are seen in this photograph. The auditorium was the scene of performances by a variety of entertainers since its opening in 1903. The building was on West Fifth Street between Main and Liberty Streets. (Courtesy of FCPL.)

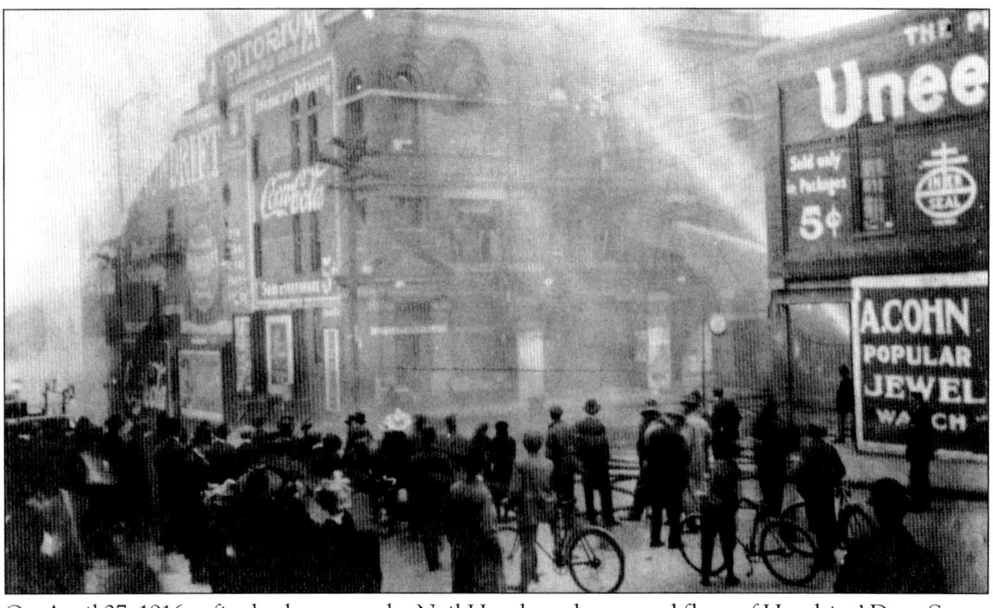

On April 27, 1916, a fire broke out at the Neil Hotel on the second floor of Hutchins' Drug Store. The drugstore and nearby Elks Auditorium, as well as several other businesses, were quickly engulfed in the fire. The fire caused roughly $90,000 worth of damage and the death of one person. This photograph shows the Elks Auditorium being sprayed with water during the fire. (Courtesy of FCPL.)

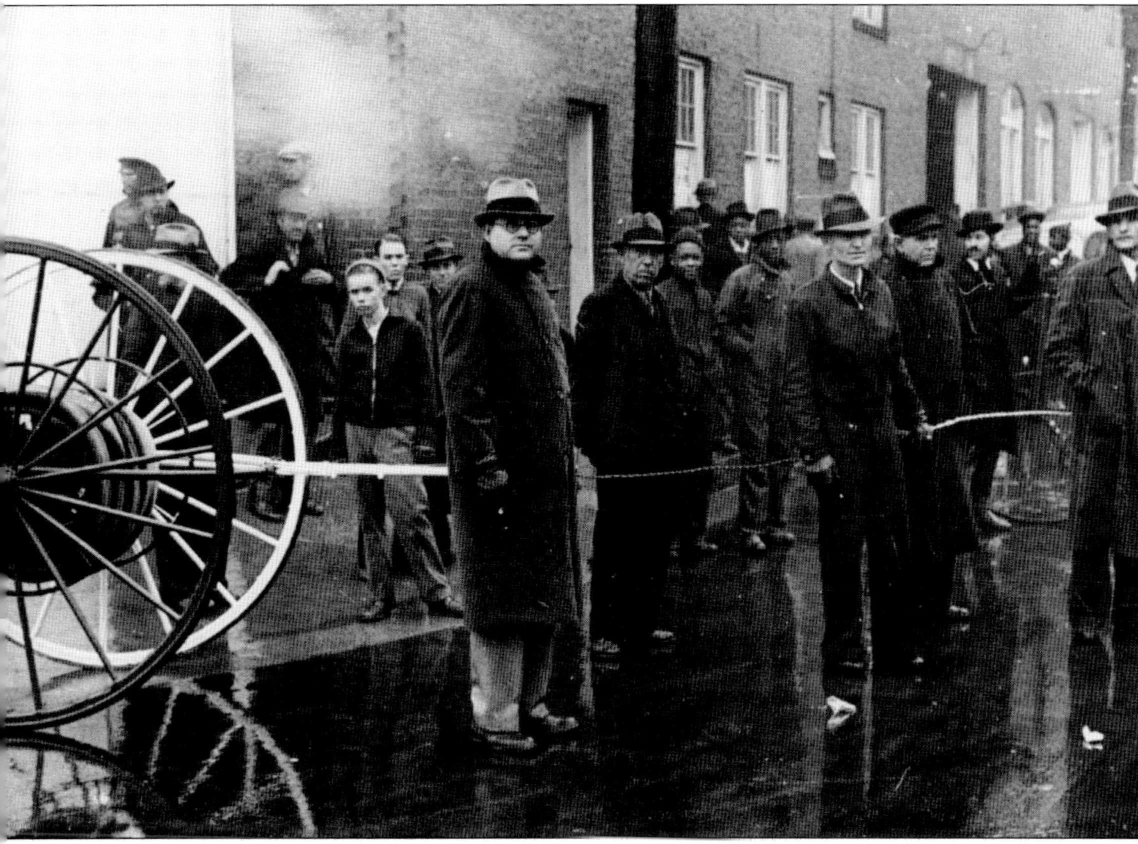

The Romance of Transportation parade was held in December 1936. The parade marked the end of the city's 46-year-old streetcar system. This photograph shows men with a reel hose used by the fire department. (Courtesy of FCPL.)

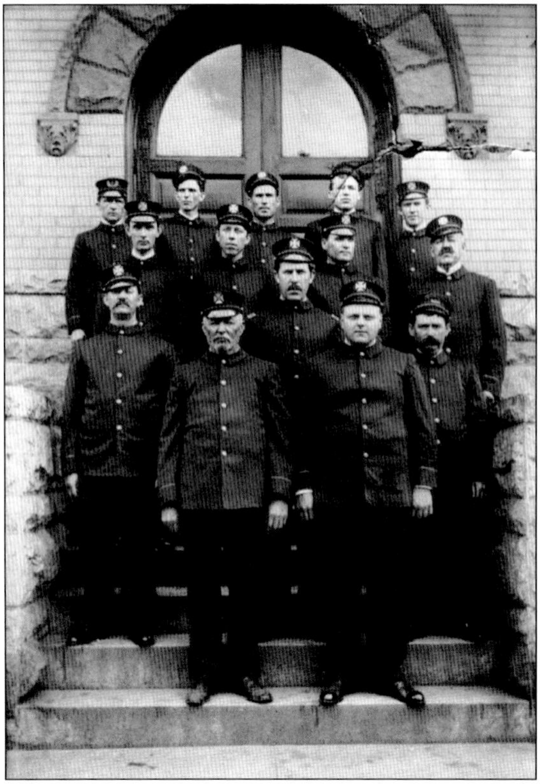

In this image of Salem Fire Department members in 1900, John H. Holmes is in the second row at right. He was the first paid fire department employee for the city of Winston-Salem. His job was to tend to the horses and keep the horse-drawn water wagon from freezing. Holmes was ultimately promoted to assistant chief and served in that capacity until his retirement on December 1, 1945, after 50 years of service. (Courtesy of the FCFHS.)

This photograph was taken in front of the Jones Hotel on North Main Street and shows the Winston Fire Department's horse-drawn pumper. From left to right are Tom Keith, Bob Shelton, John H. Holmes, Jule Stith, Tom Graham, and Red Powell (little boy). (Courtesy of FCPL.)

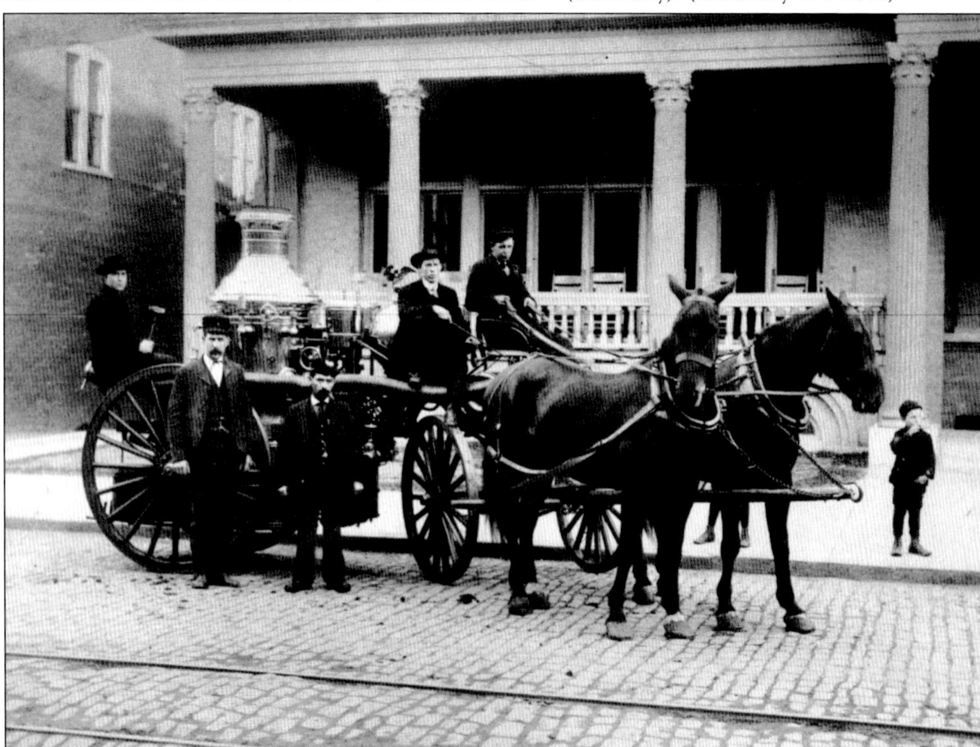

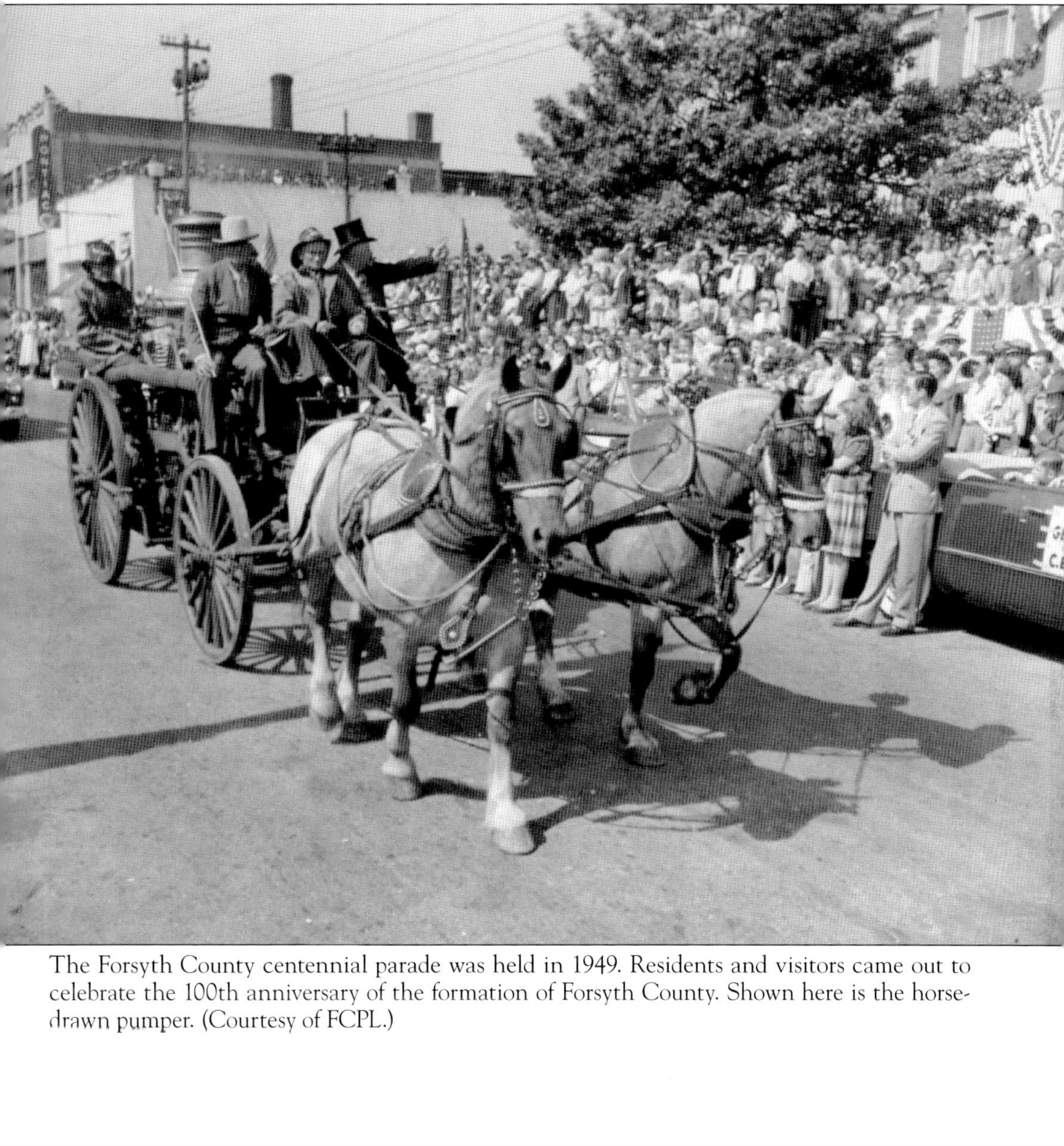

The Forsyth County centennial parade was held in 1949. Residents and visitors came out to celebrate the 100th anniversary of the formation of Forsyth County. Shown here is the horse-drawn pumper. (Courtesy of FCPL.)

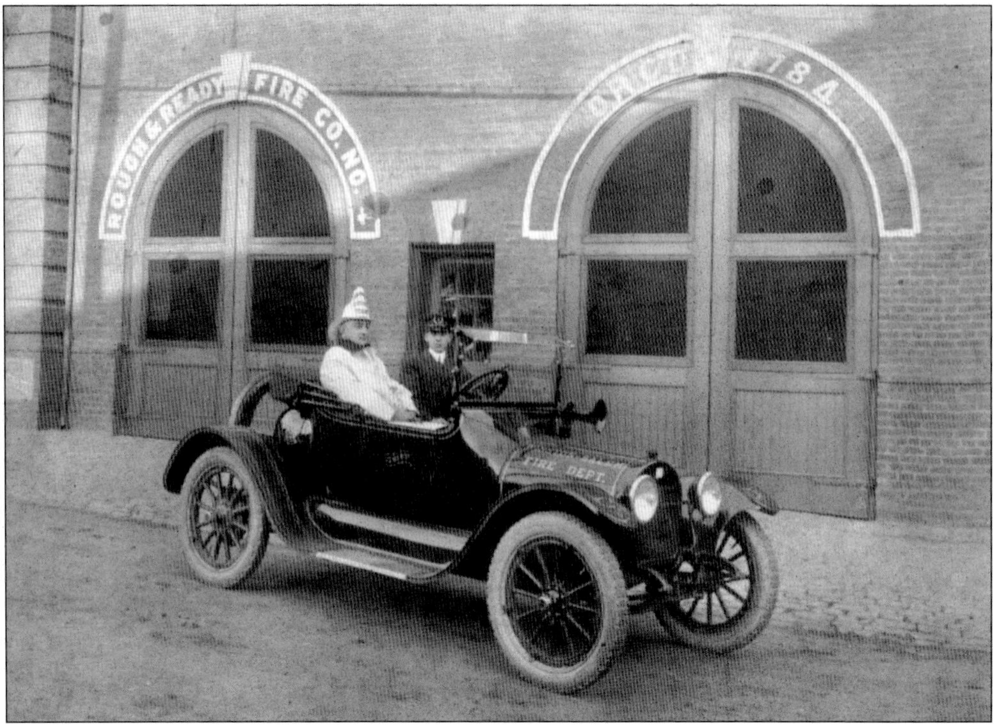

Chief H.E. Nissen and his driver, Jinkins, are seen sitting in the city's new American LaFrance chief's car. The chief and his driver were killed in 1932 responding to an alarm in this very car. (Courtesy of Winston-Salem Fire and Rescue Historical Society.)

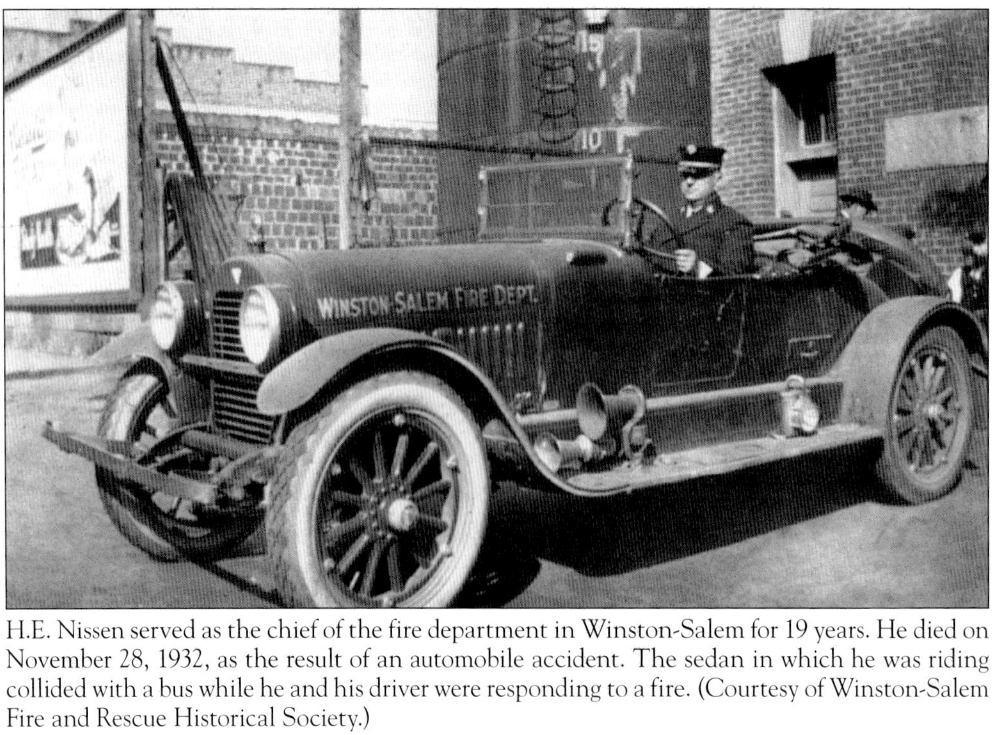

H.E. Nissen served as the chief of the fire department in Winston-Salem for 19 years. He died on November 28, 1932, as the result of an automobile accident. The sedan in which he was riding collided with a bus while he and his driver were responding to a fire. (Courtesy of Winston-Salem Fire and Rescue Historical Society.)

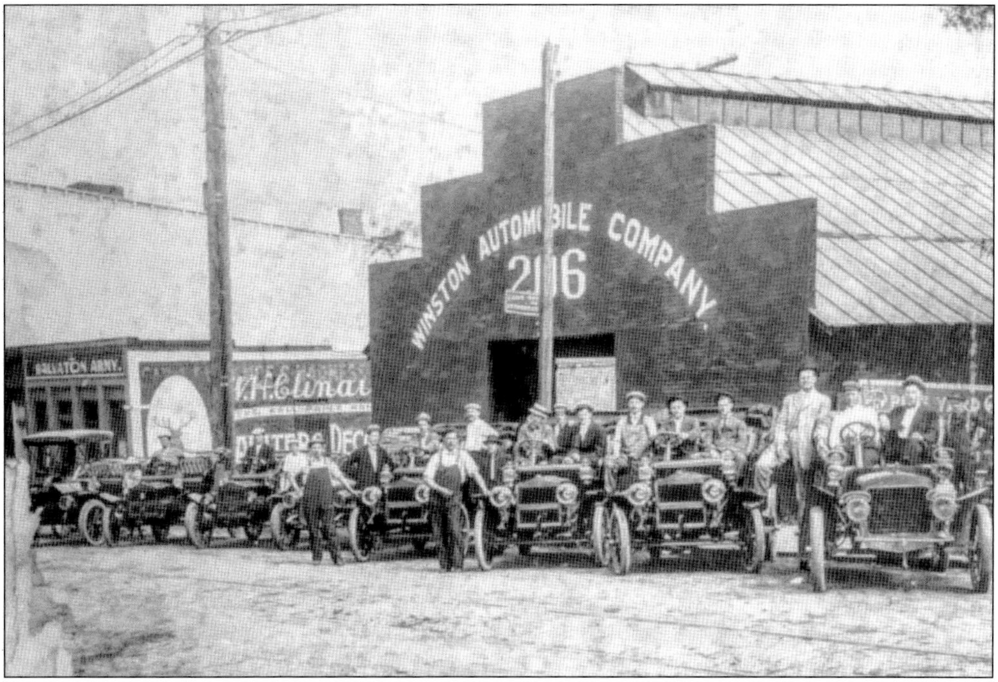

This photograph was taken in 1908 and shows the growth of businesses in the city. The Winston Automobile Company was at 206 West Fourth Street. The Salvation Army is on the left. (Courtesy of FCFHS.)

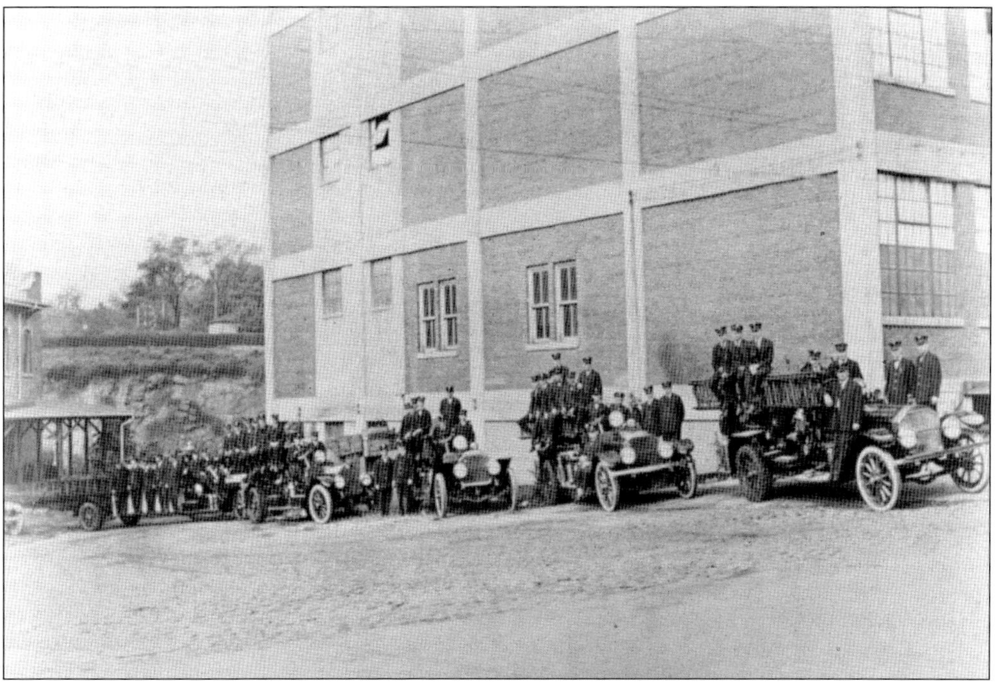

The Winston-Salem Fire Department is shown in 1920 beside what later became Downtown Middle School. Five fire engines can be seen with firemen seated on or standing next to them. (Courtesy of Winston-Salem Fire and Rescue Historical Society.)

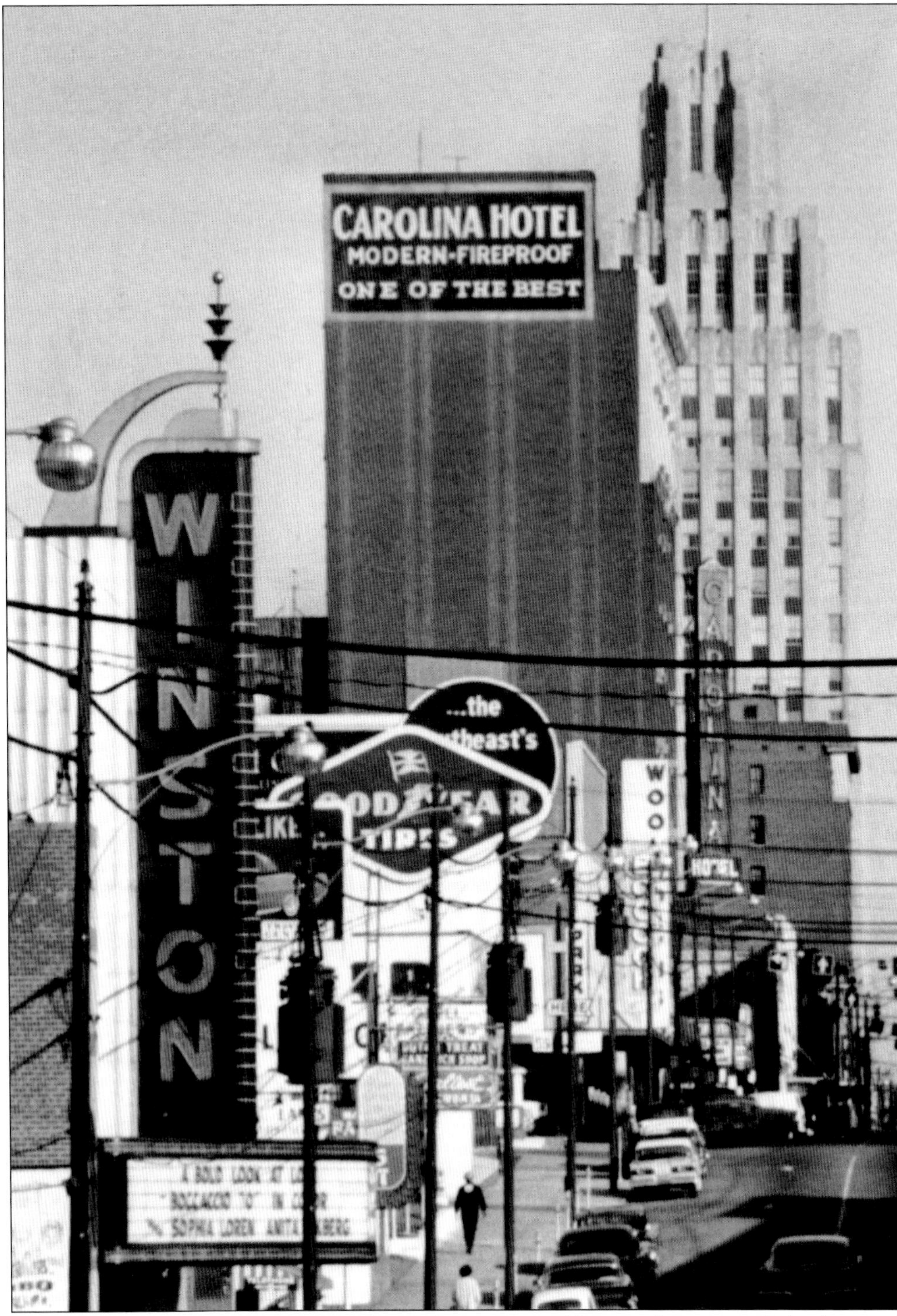

The Reynolds Building, Carolina Hotel, and the Nissen Building can be seen in the background. The town of Winston-Salem was growing both in population and in industry. (Courtesy of FCPL.)

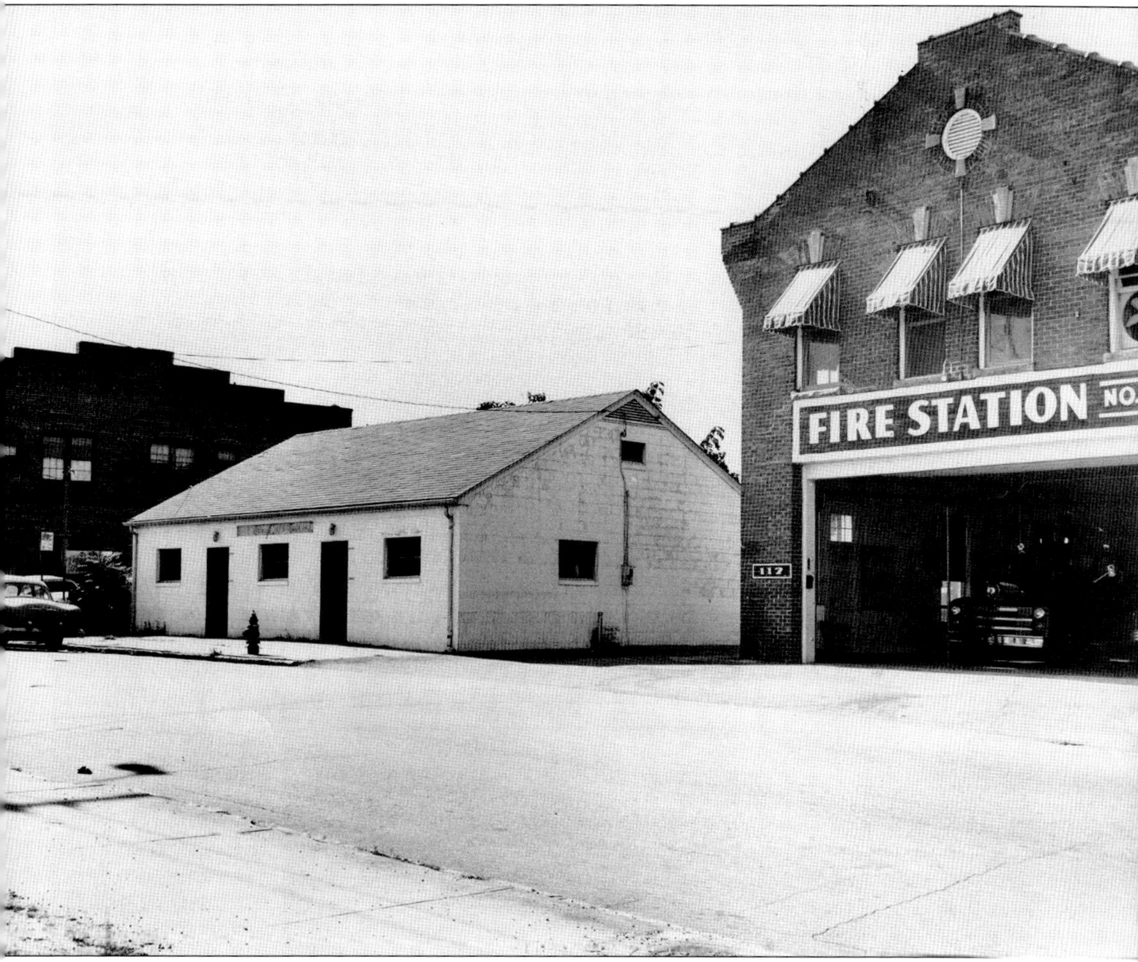

Winston-Salem Fire Station No. 1 was on Eighth Street. The city-owned public restrooms are seen here beside the fire station. The station housed the department until 1976. (Courtesy of FCPL.)

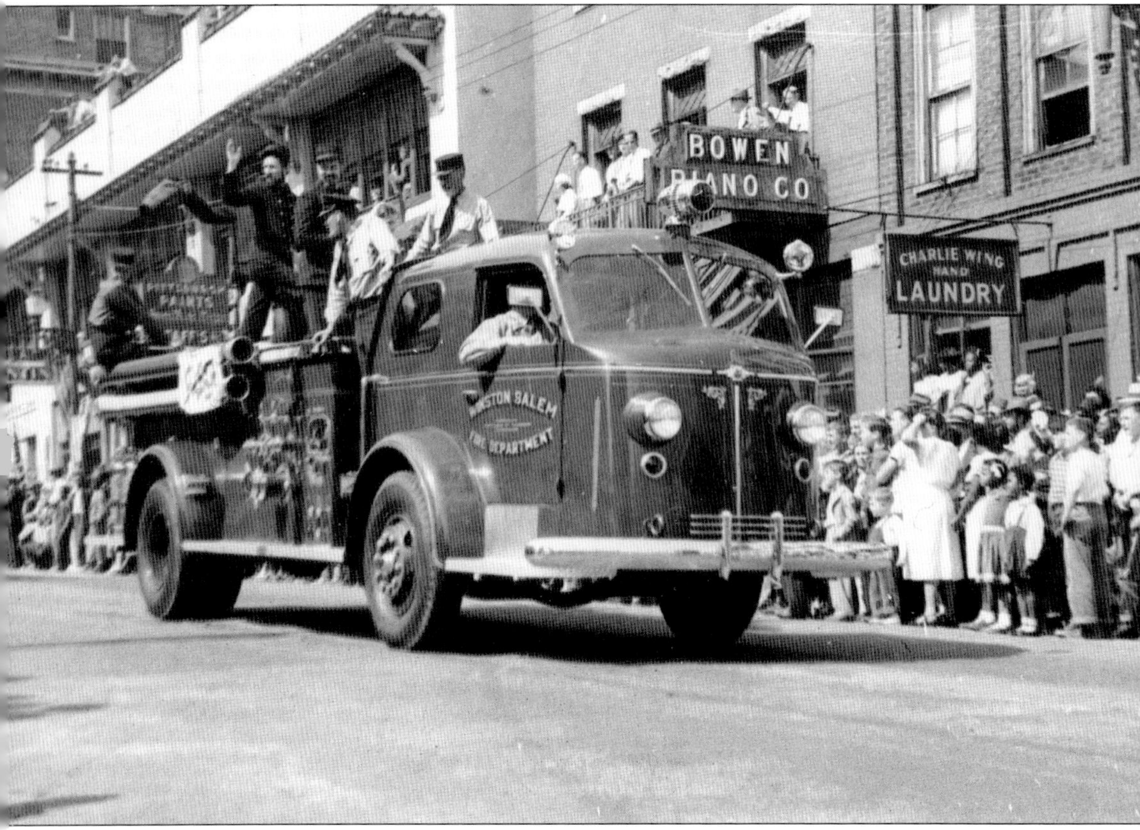

The Fire Prevention Week parade was held in 1955. The parade went through downtown Winston-Salem and displayed the city's apparatus and personnel. This photograph shows fire trucks on West Fourth Street. (Courtesy of FCPL.)

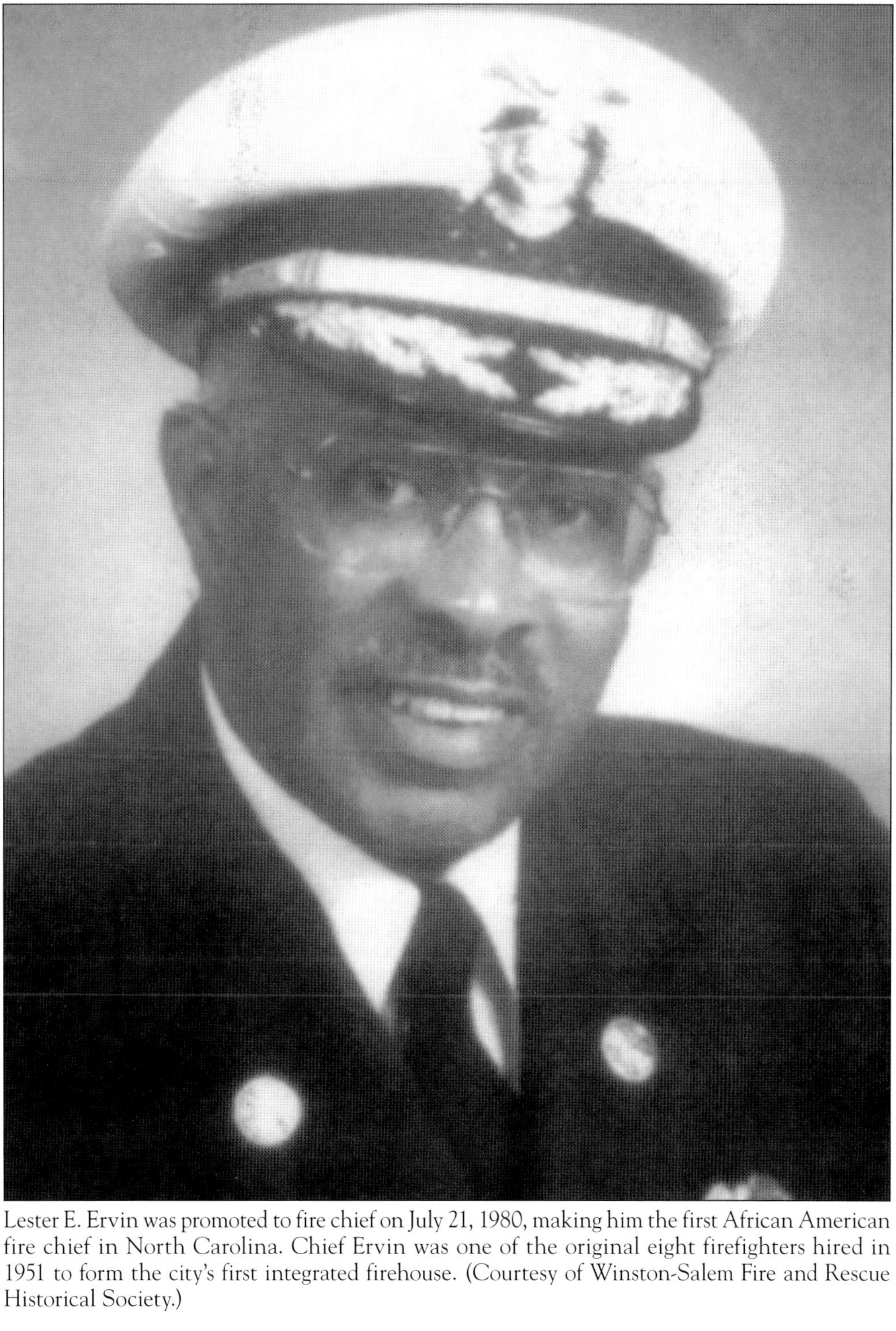

Lester E. Ervin was promoted to fire chief on July 21, 1980, making him the first African American fire chief in North Carolina. Chief Ervin was one of the original eight firefighters hired in 1951 to form the city's first integrated firehouse. (Courtesy of Winston-Salem Fire and Rescue Historical Society.)

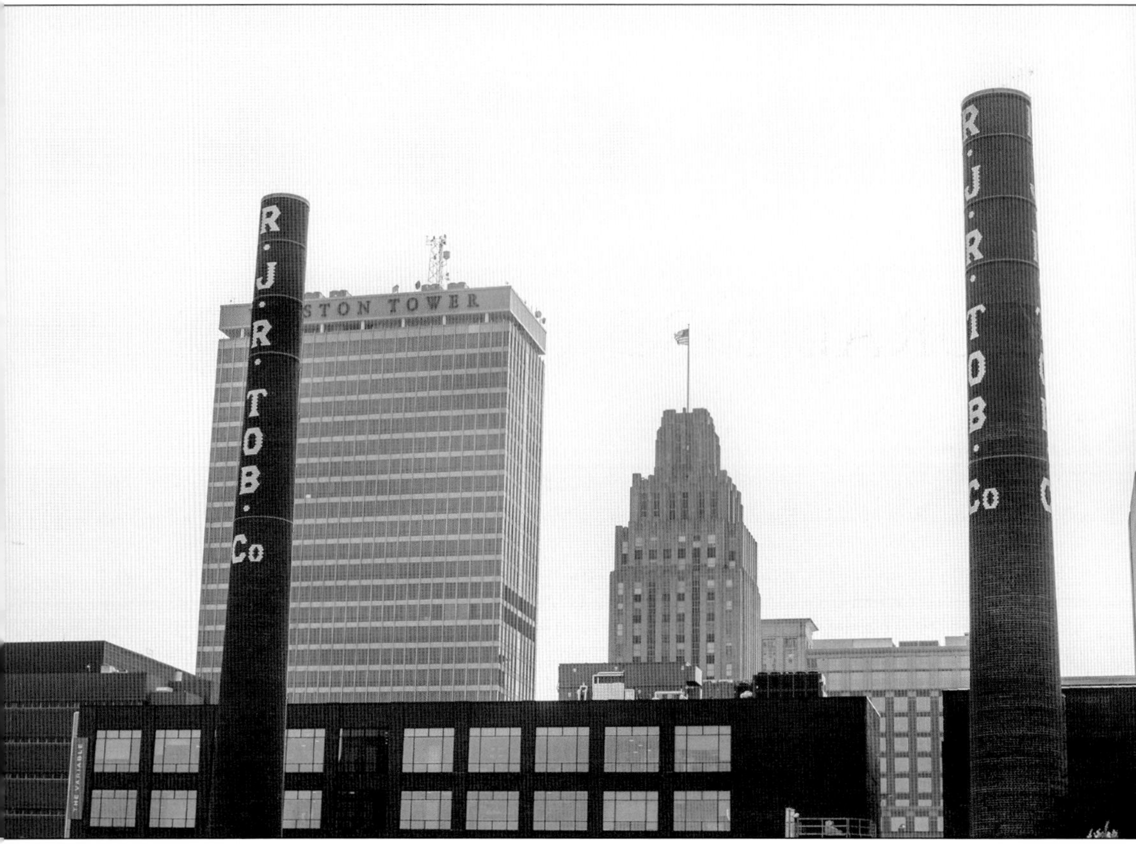

Winston-Salem's distinctive skyline features the Reynolds Building and smokestacks from the original Richard Joshua Reynolds Tobacco Company factory. As industry became more productive, it enticed more people to move to the area. (Courtesy of Sylvia Elizabeth Photography.)

Two

Rural Fire Protection

The industries of Forsyth County grew and brought growth to the county. Rural homes, farms, churches, and roadways were the new frontier for fire protection. Agriculture drove North Carolina's economy, with over 50 percent of adult males reporting that they were farmers. Tobacco was a major commodity, able to be grown in many different soil types and climates. Fortunately, Forsyth County farmers had some of the largest tobacco businesses in their backyards. The Richard Joshua Reynolds Tobacco Company produced 25 percent of the nation's tobacco. The vast majority of Forsyth County farmers sold their tobacco at auction houses in Winston and Salem. There were smaller tobacco factories in Bethania, Kernersville, Old Town, and Walkertown, which supplied farmers with several market locations. Tobacco drove the state economy for many years, and Forsyth County shared much credit in sustaining it.

The town of Kernersville established its fire department in November 1923. Kernersville later founded the first combination fire department in the area with both volunteer and paid staff. In 1964, the town hired two firefighters who worked 24-hour shifts. The Adam Millis Plant on Main Street in Kernersville burned down on March 26, 1981, displacing over 500 people. It was the largest financial fire loss in the state at the time.

The next town to establish a plan for fire protection was Rural Hall, due to continued growth. The only fire protection in Rural Hall in 1941 came from a large-wheeled fire extinguisher purchased by the town for area merchants. In 1949, a two-story building was designated to store fire equipment such as hoses, ladders, and extinguishers. The town later purchased a 1941 Plymouth truck for $500, with an additional $25 invested to paint it "fire engine red."

In the 1940s, schoolhouses in Mineral Springs and Lewisville caught fire, causing extensive damage. Numerous homes and churches also suffered damage, illustrating the need for more rural protection. In 1941, the Forsyth County School Board purchased firefighting equipment and began training high school seniors. These exercises were led by members of the Winston-Salem Fire Department. As communities grew, fire wardens were established as part of civil defense during World War II.

Several fires made headlines and spurred concern over the lack of a formal rural fire protection plan. This helped pave the way for volunteer fire departments. To help establish a new department, Forsyth County offered financial aid for the maintenance and operation of fire apparatus: $50 for the first truck and $15 for the second, plus an annual fee of $100. Each department was also allowed to borrow $1,000 to be repaid in 10 years without interest. A total of $300 was given to each department for the purchase of equipment.

The development of the fog nozzle was crucial to rural fire service. Chief Lloyd Layman of Parkersburg, West Virginia, introduced the phrase "the indirect method of attack" to suppress interior building fires using the heat-absorbing properties of steam, produced in great quantities by fog streams. The theory of indirect fire attacks was based on Navy and Coast Guard experience battling fires on ships during World War II. With this method, droplets cover a greater surface area for a faster rate of heat absorption, transforming into steam and smothering the fire.

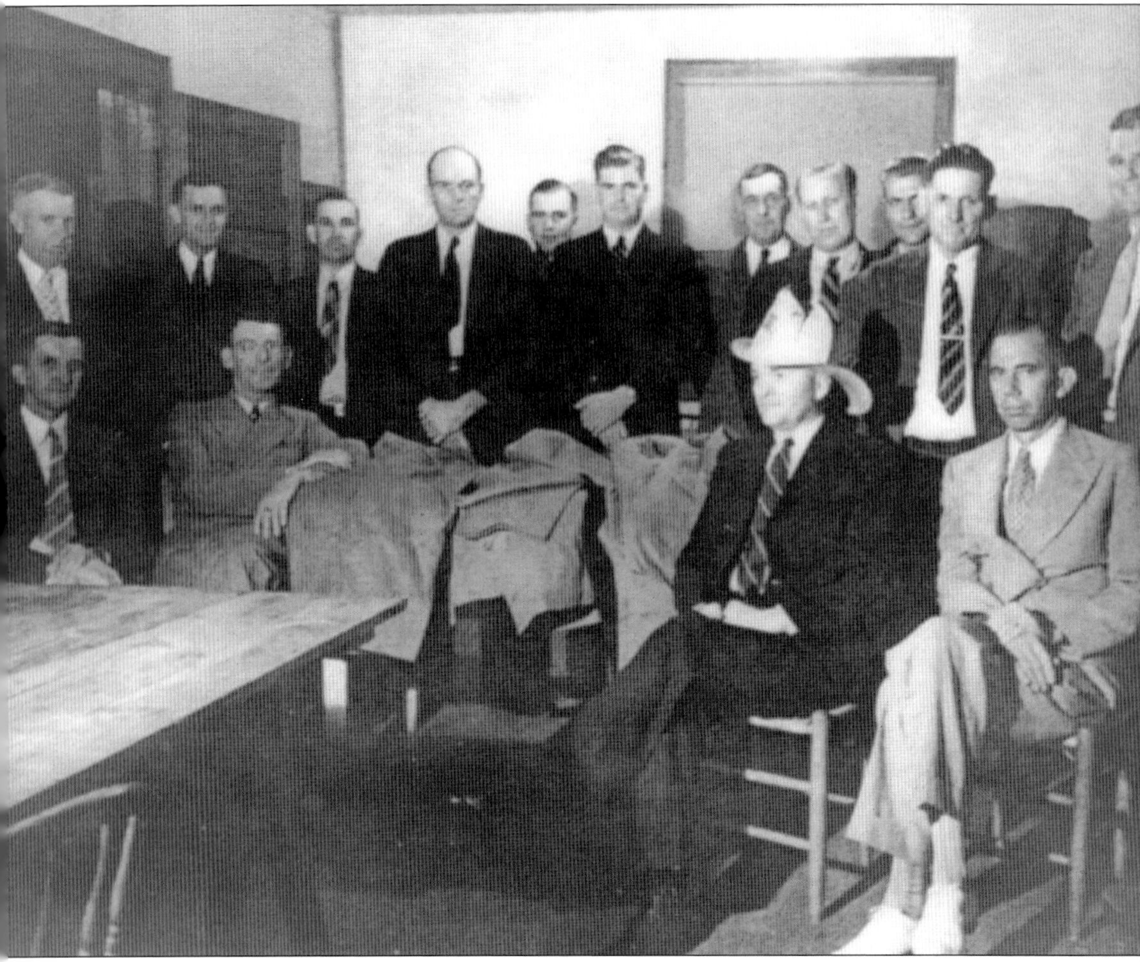

Several volunteer firemen gathered at Chief Ned R. Stuart's automobile dealership to discuss fire training objectives and tactics for the Kernersville Volunteer Fire Department in 1926. Chief Stuart is seated wearing a white helmet, with H. Kirk seated beside him. Stuart served as fire chief from 1923 to 1963. The department was all volunteer until December 1964, when the town hired two full-time firefighters who worked 24 hours on and 24 hours off. The department hired its first full-time fire chief in 1970. Standing in the back row are, from left to right, B. Stafford, unidentified, H. Albert, C. Stuart, C. Simpson, B. Linville, A.C. Ballard, P. Linville, G. Whicker, W. Parrish, and H. Linville. The two men seated at left are unidentified. (Courtesy of FCFHS.)

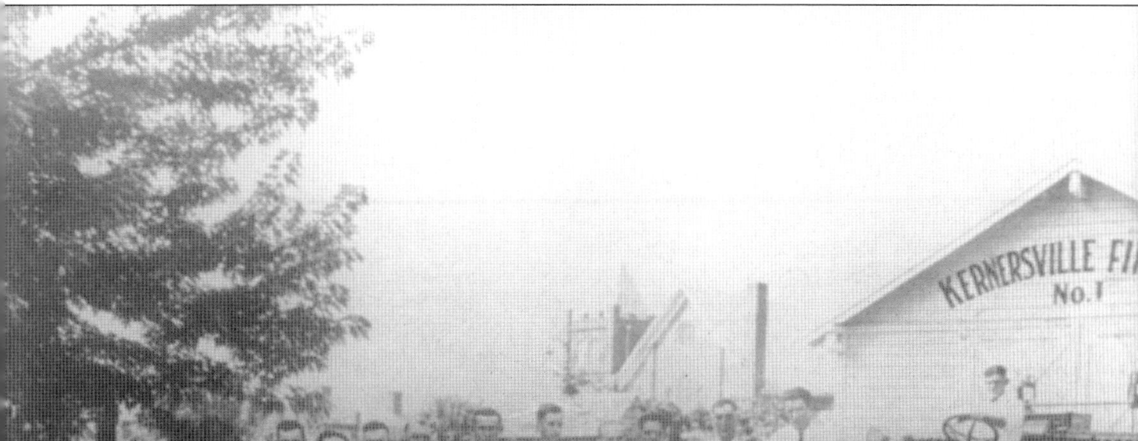

This was the Kernersville Fire Department in 1925. Ned R. Stuart organized the first volunteer fire department, and that same year, the town of Kernersville purchased a 1923 American LaFrance fire engine, pictured here, for $6,500. Prior to this, bucket brigades and wells were the best fire service the town had to offer. This truck was housed in Stuart's garage until the department's building was finished on East Mountain Street in 1928. From left to right are Jess Byers, Bern Stafford, "Little" Tony Ballard, D. Phipps, Herb Stafford, Warner Jesse, Arthur Ballard, Johnny Nelson, "Red" ?, Sam F. Vance, Vern Harrison, June Woollave, Ned Stuart, George Perry, Charlie Dillon, "Big" Tony Ballard, and Wade Parrish. (Courtesy of Kernersville Museum.)

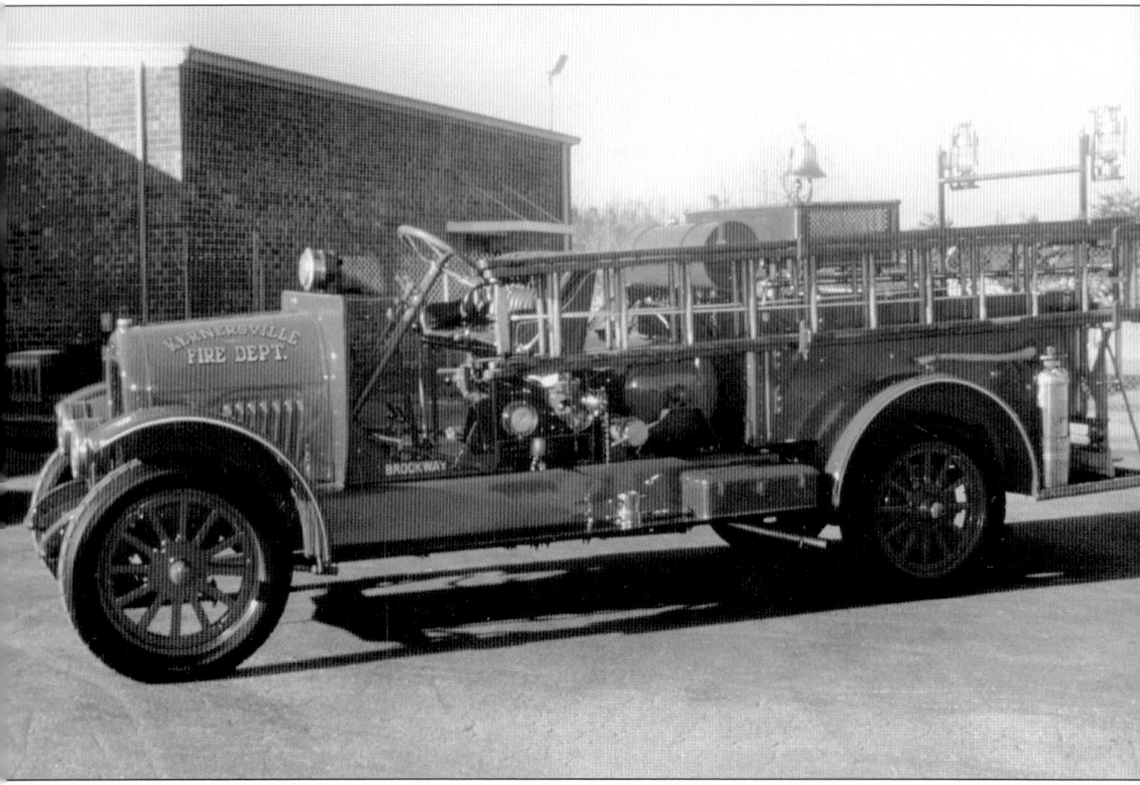

The Kernersville Fire Department purchased this 1926 American LaFrance fire engine from the Pittsburgh area in 1940 as a dependable spare. The engine was later sold in 1968. (Courtesy of FCFHS.)

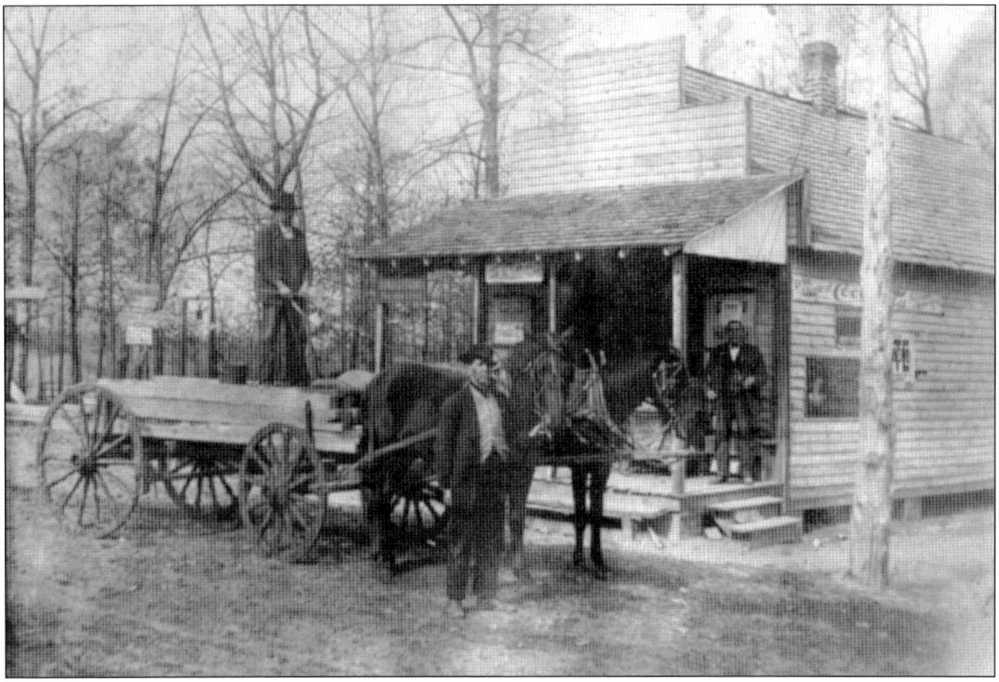

The Westmorland general store was in rural Forsyth County and was owned by John Wiley Westmorland, standing on the porch. Thomas Walker is driving the horses, and Charles Westmoreland, son of John, is standing beside them. (Courtesy of FCPL.)

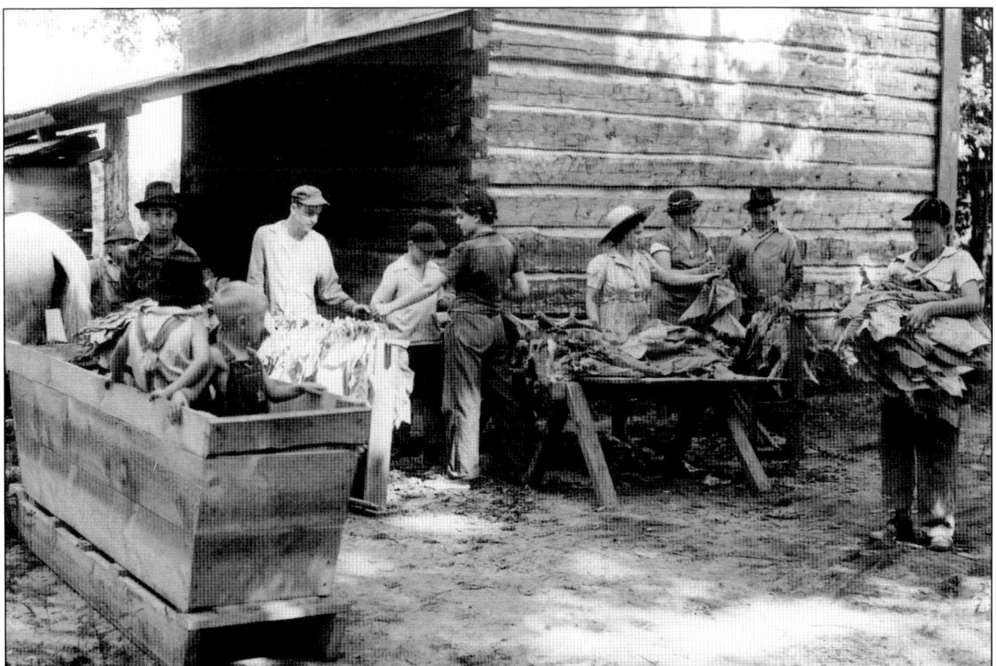

A rural farm family is shown here in 1941 tying tobacco to poles for curing. Much of the nation's tobacco was farmed in Forsyth County on family farms just like this. Fire protection for these farmers was essential. (Courtesy FCPL.)

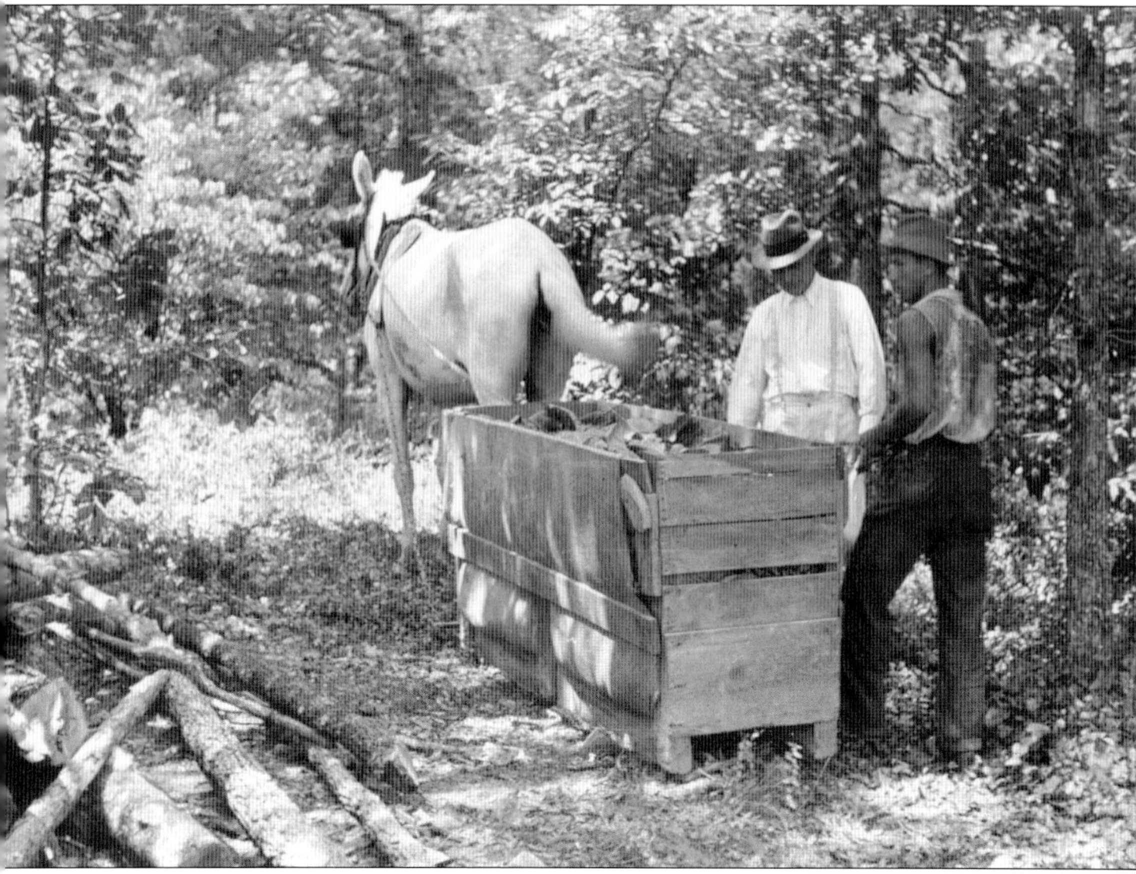

Sam Fulp (left) and Robert Shutt are transporting a sled holding tobacco leaves. Shutt was a local tobacco farmer, and Fulp ran a general store in Forsyth County. (Courtesy of FCPL.)

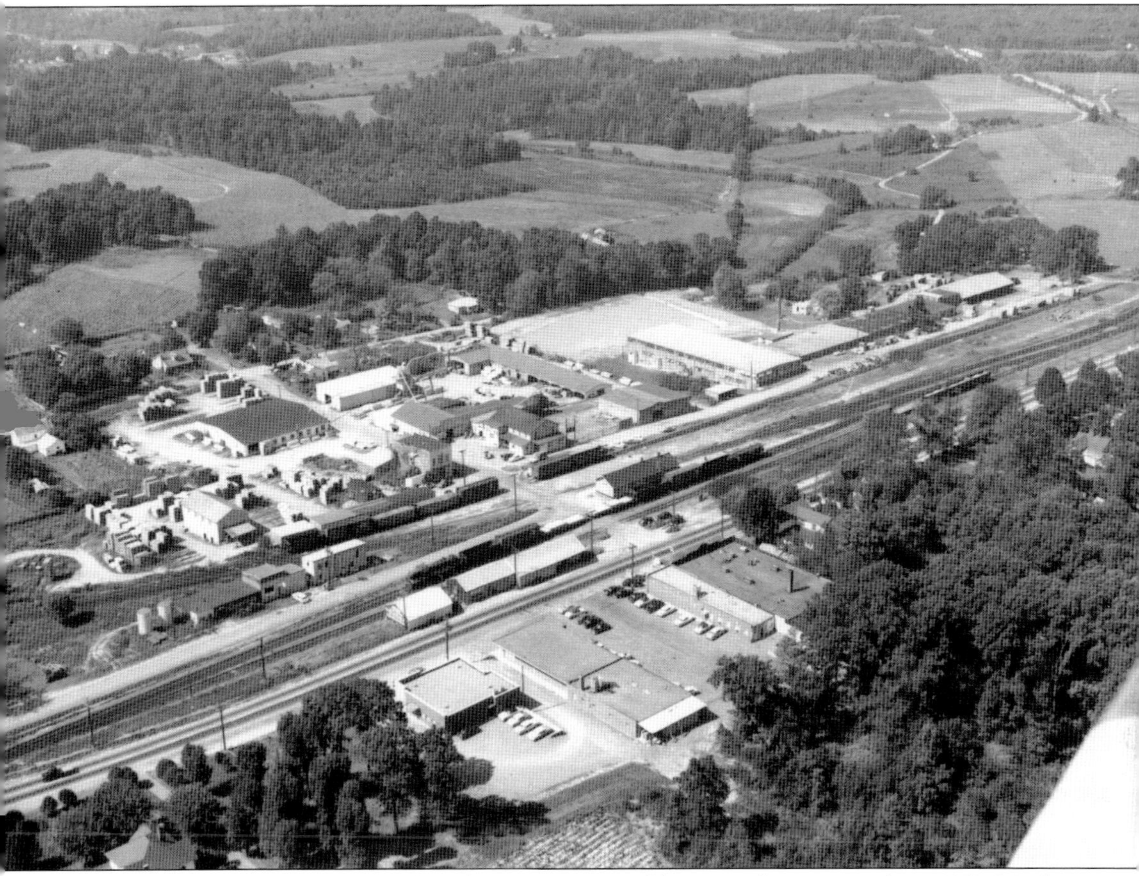

The town of Rural Hall was developed in 1907. Pictured here are the train yards in 1958. Much of the city and county's success was only gainful because of the railroad. (Courtesy of FCPL.)

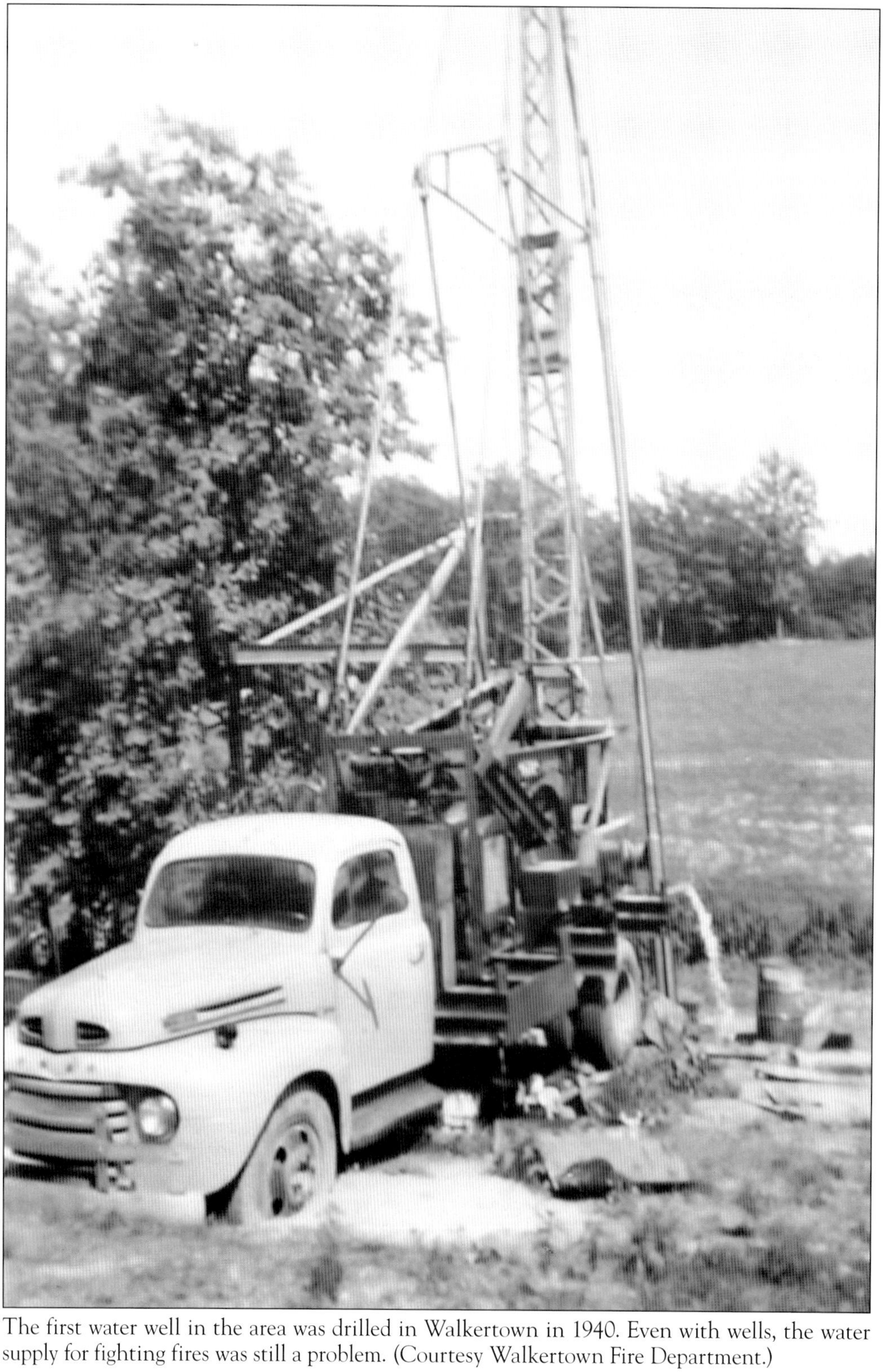

The first water well in the area was drilled in Walkertown in 1940. Even with wells, the water supply for fighting fires was still a problem. (Courtesy Walkertown Fire Department.)

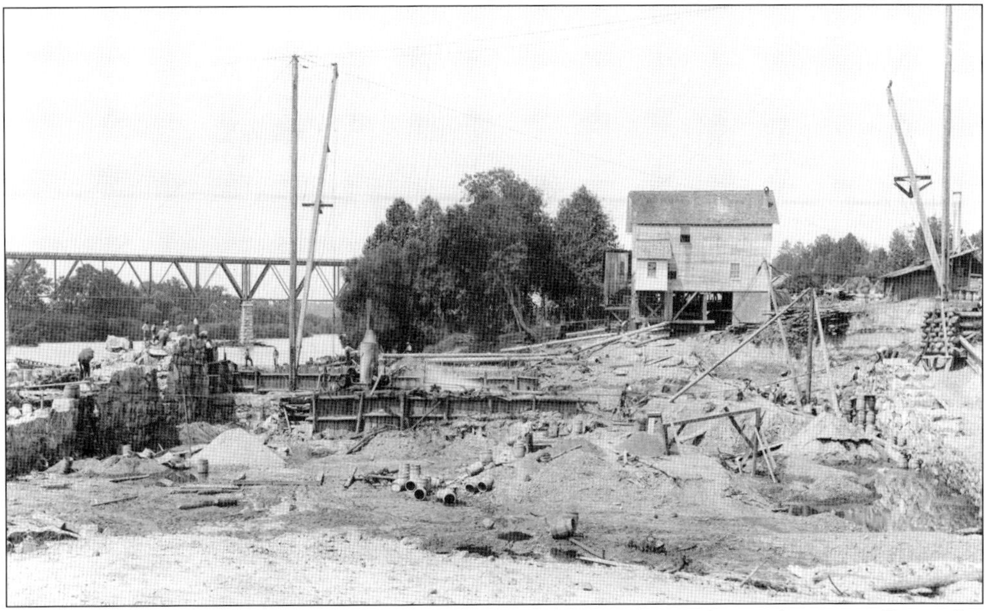

Idol's Dam was built in 1898 on the Yadkin River in Clemmons, North Carolina. A mill can also be seen in this photograph. (Courtesy of FCPL.)

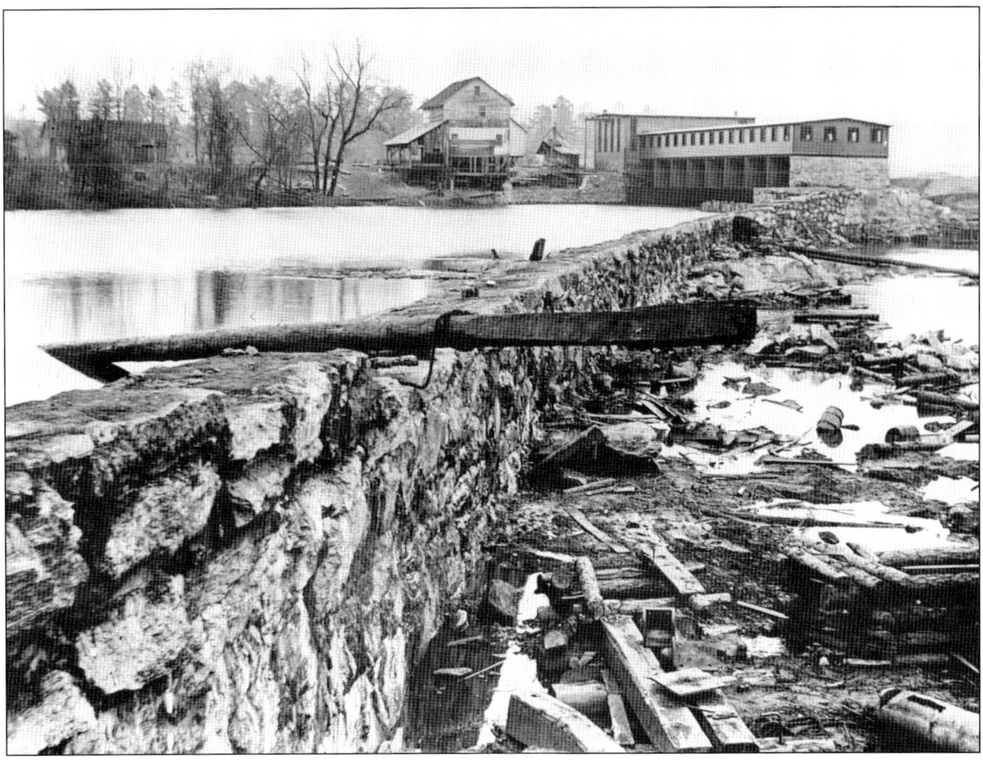

The Idol's Dam power station was the first commercial hydroelectric generating station in North Carolina. Located just west of Clemmons on the Yadkin River, it provided long-distance transmission of alternating current. Thomas Edison, the famous inventor, was an early investor in the project. (Courtesy of FCPL.)

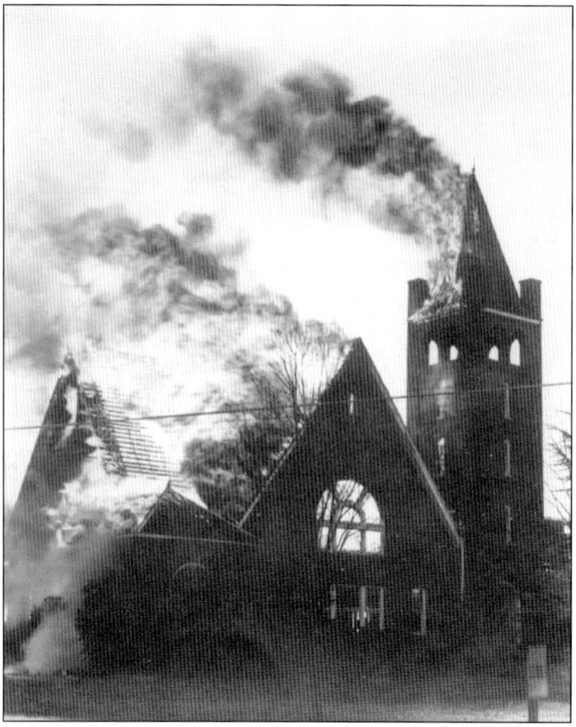

A large fire caused by a heating stove destroyed Loves Methodist Church in 1949. Shortly after, the Walkertown Volunteer Fire Department was formed. (Both, courtesy of FCFHS.)

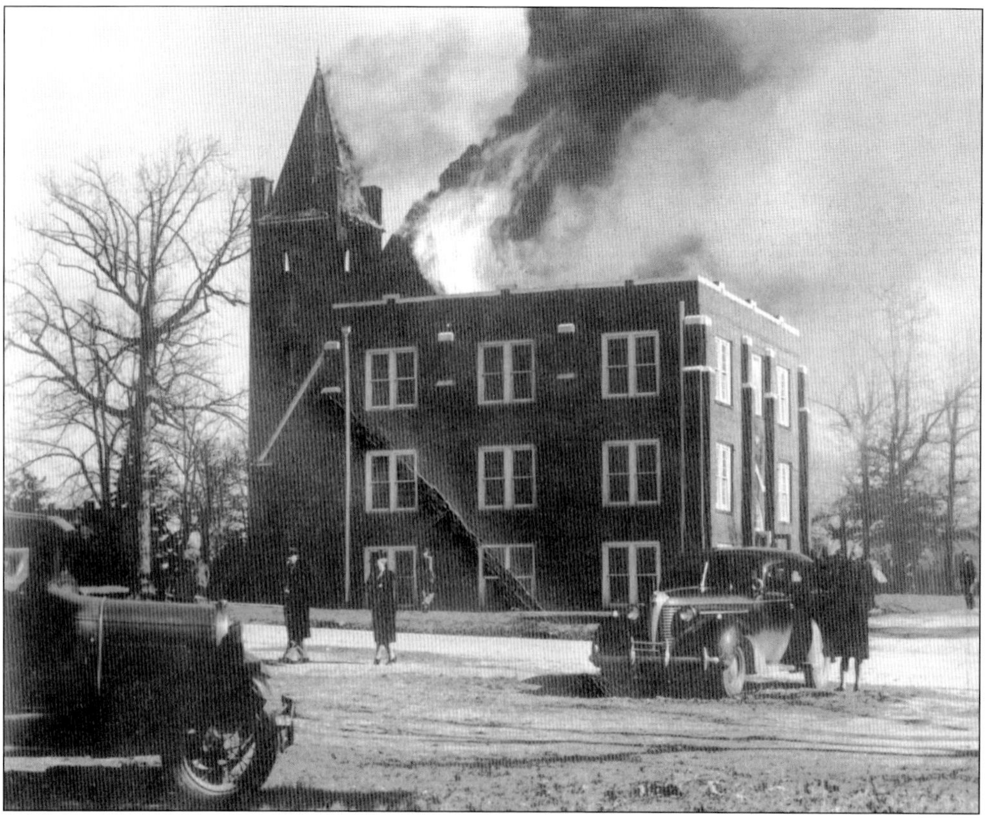

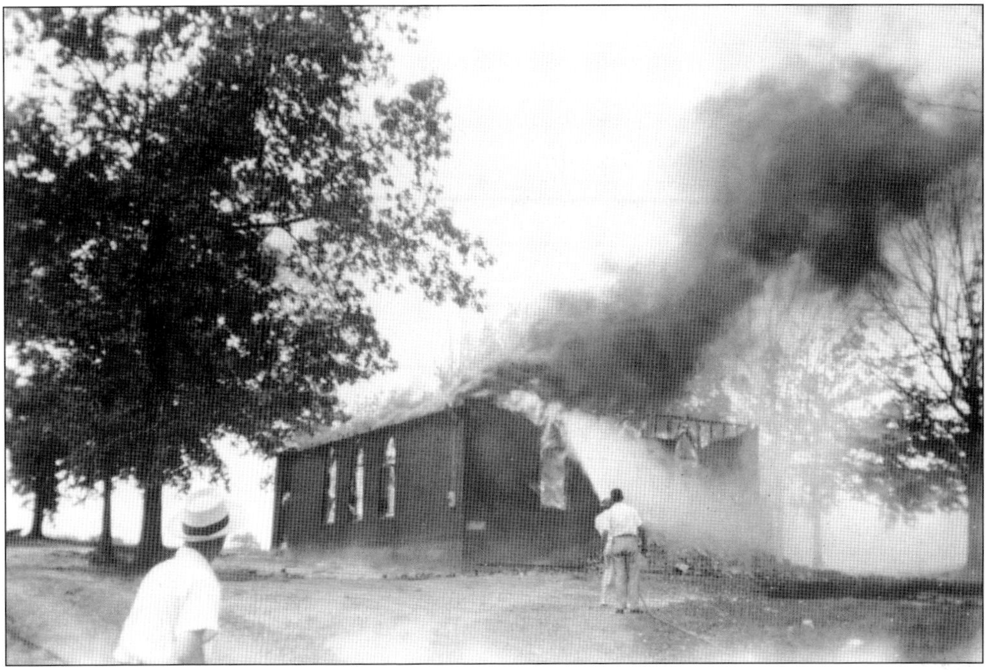

The Oak Grove Baptist Church burned in 1950. The Walkertown Volunteer Fire Department was quick to respond, but the fire had grown out of control due to the building construction and weather. (Courtesy of FCFHS.)

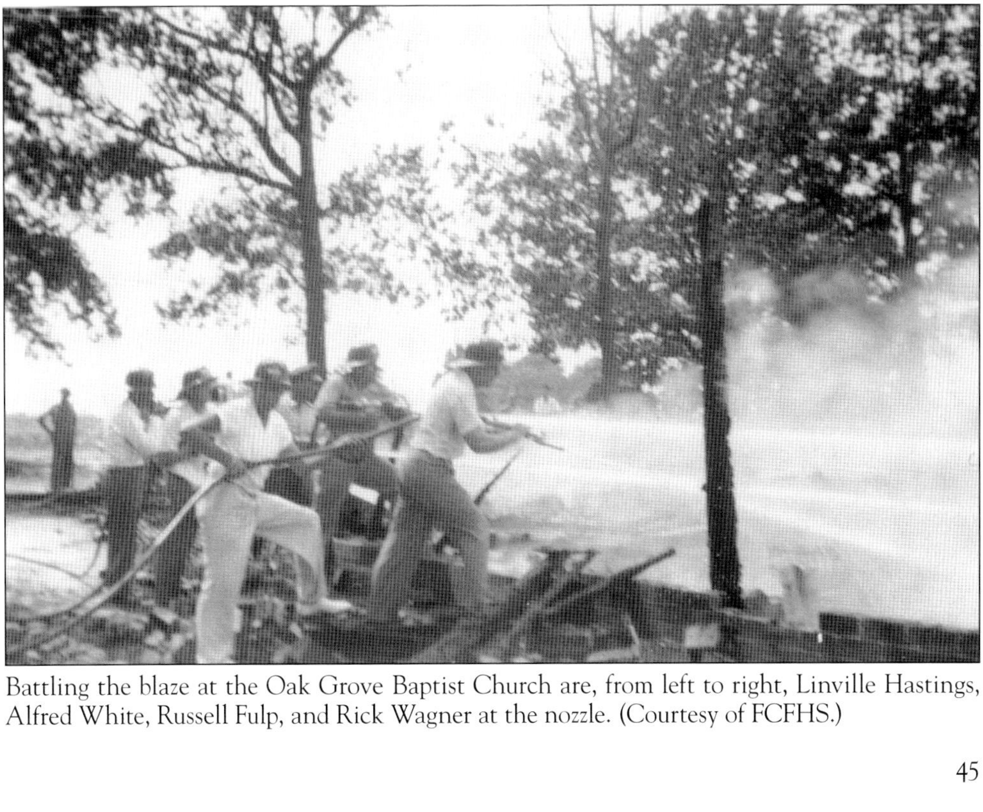

Battling the blaze at the Oak Grove Baptist Church are, from left to right, Linville Hastings, Alfred White, Russell Fulp, and Rick Wagner at the nozzle. (Courtesy of FCFHS.)

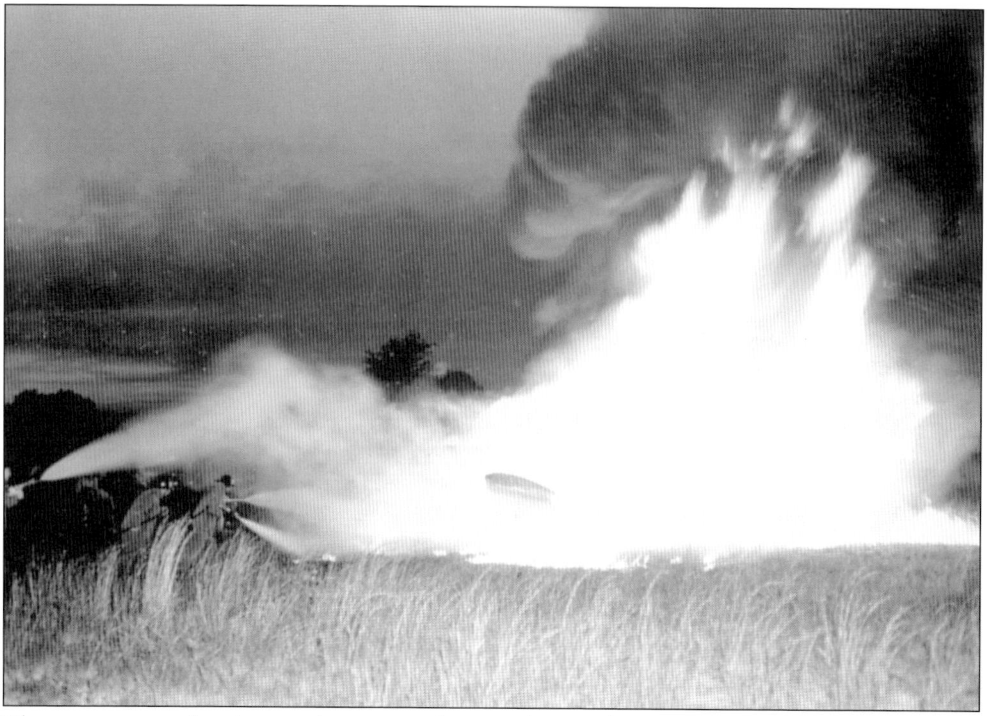

There were many farming machinery accidents as the population grew in rural Forsyth County. Pictured is a piece of equipment that caught fire in 1945. (Courtesy of FCFHS.)

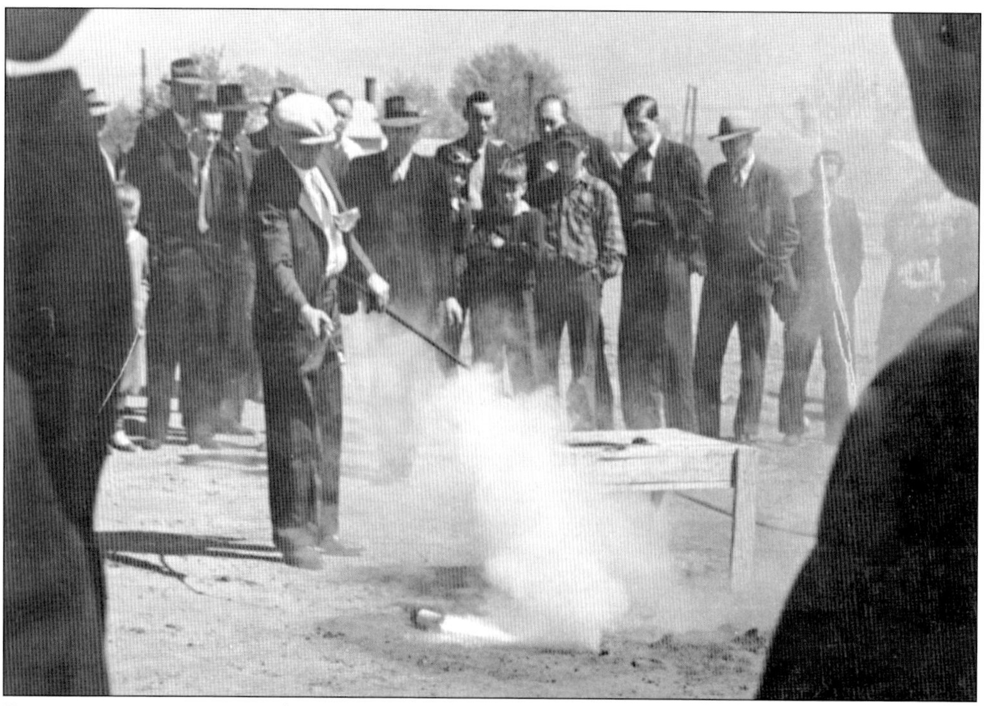

During a training event, Chief Ned R. Stuart demonstrates the dangers of white phosphorus. Chief Stuart was a civil defense warden during World War II. (Courtesy of FCFHS.)

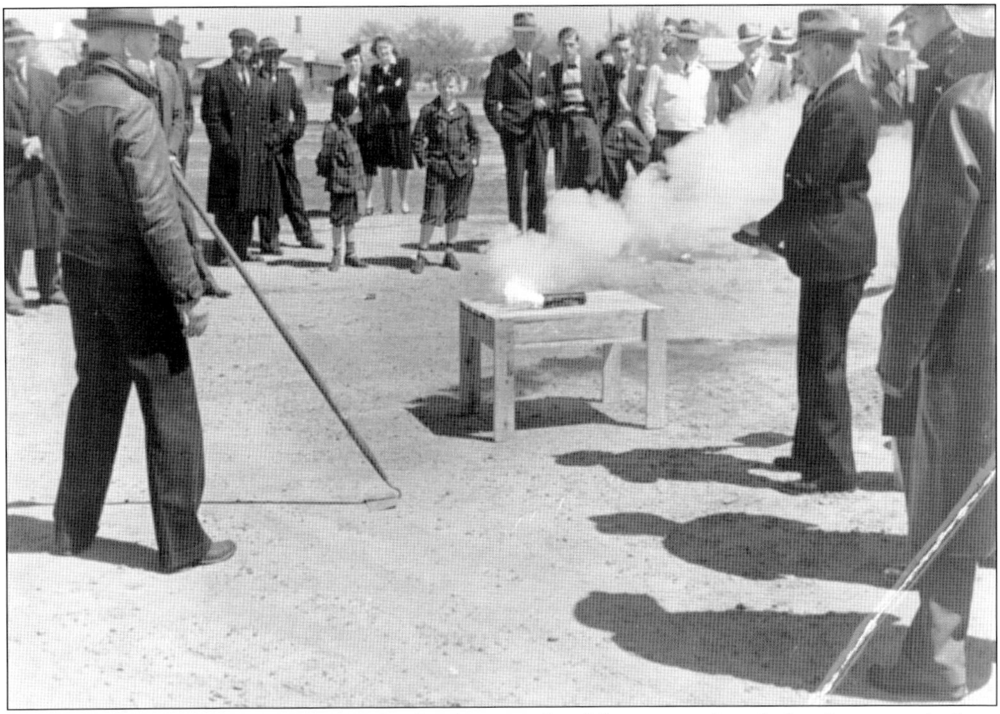

The public is seen here learning the fundamentals of fire control and the dangers of white phosphorus in 1945. (Courtesy of FCFHS.)

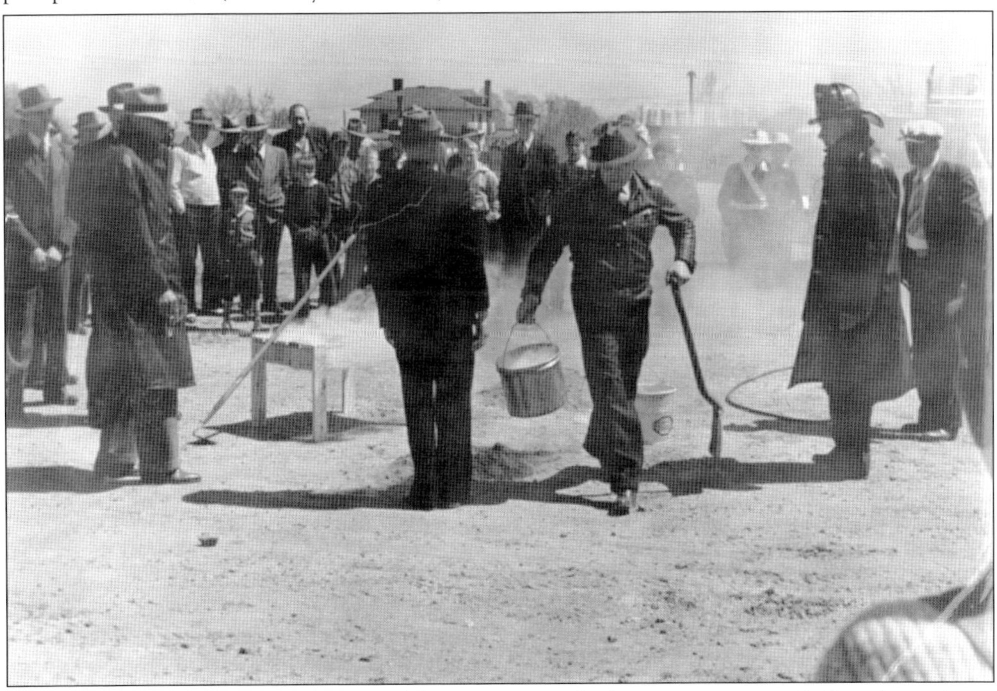

A fire and hose demonstration at a public training and education event in 1945 is shown here. Much of the training was based on the experiences of soldiers, sailors, and marines during World War II. (Courtesy of FCFHS.)

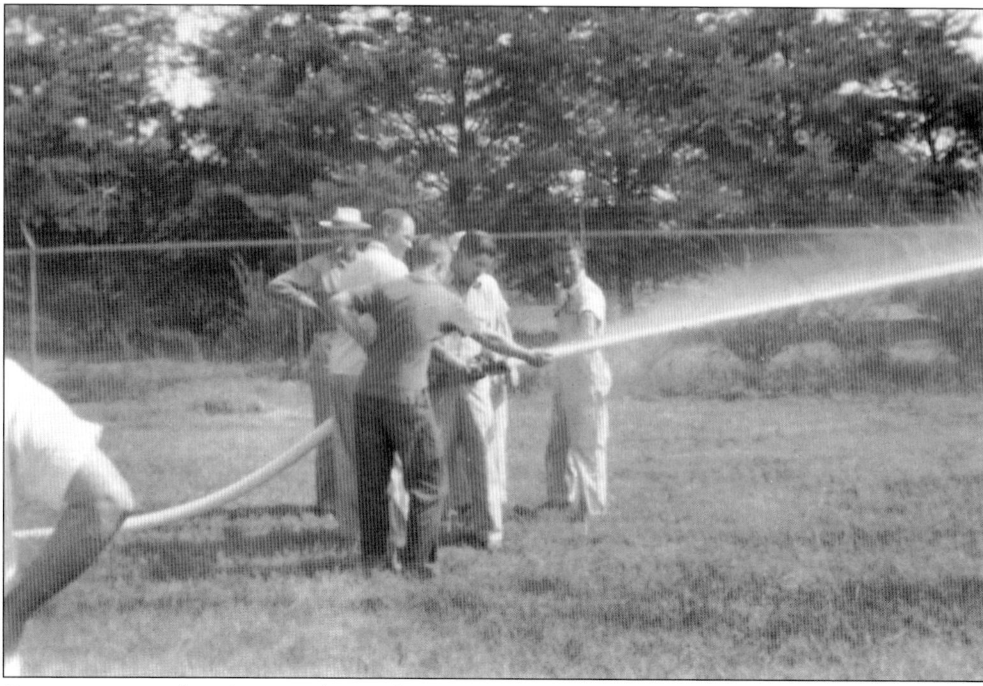

Community members from rural Forsyth County are seen here learning how to deploy and operate new fire hoses and attack nozzles. (Courtesy of FCFHS.)

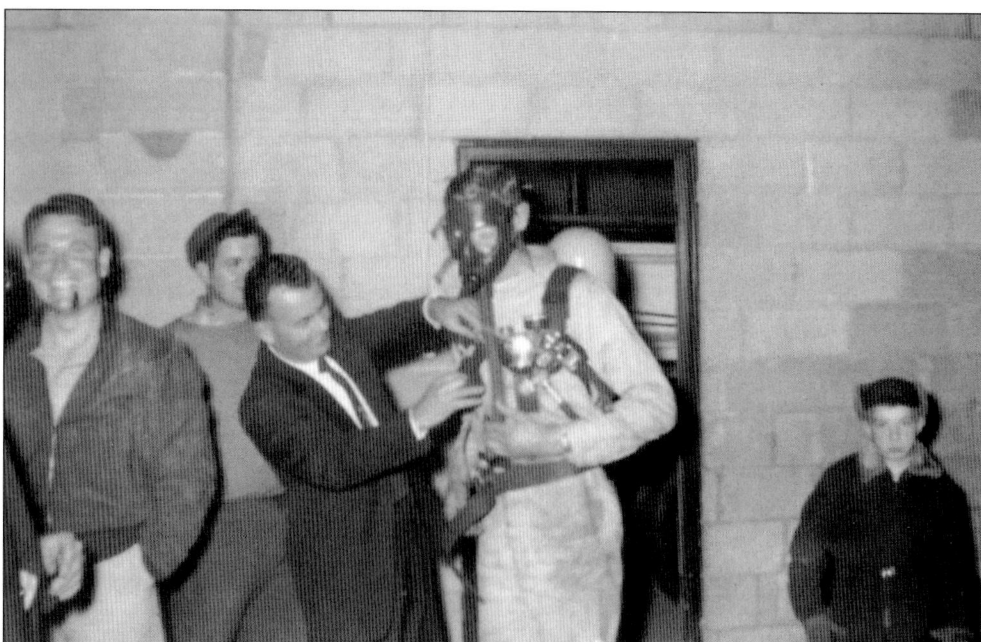

Here, members of the City View Fire Department are learning how to use the new self-contained breathing apparatus (SCBA) equipment and also how to work while wearing it. SCBA allowed firefighters to operate inside structures without breathing deadly gasses. (Courtesy of FCFHS.)

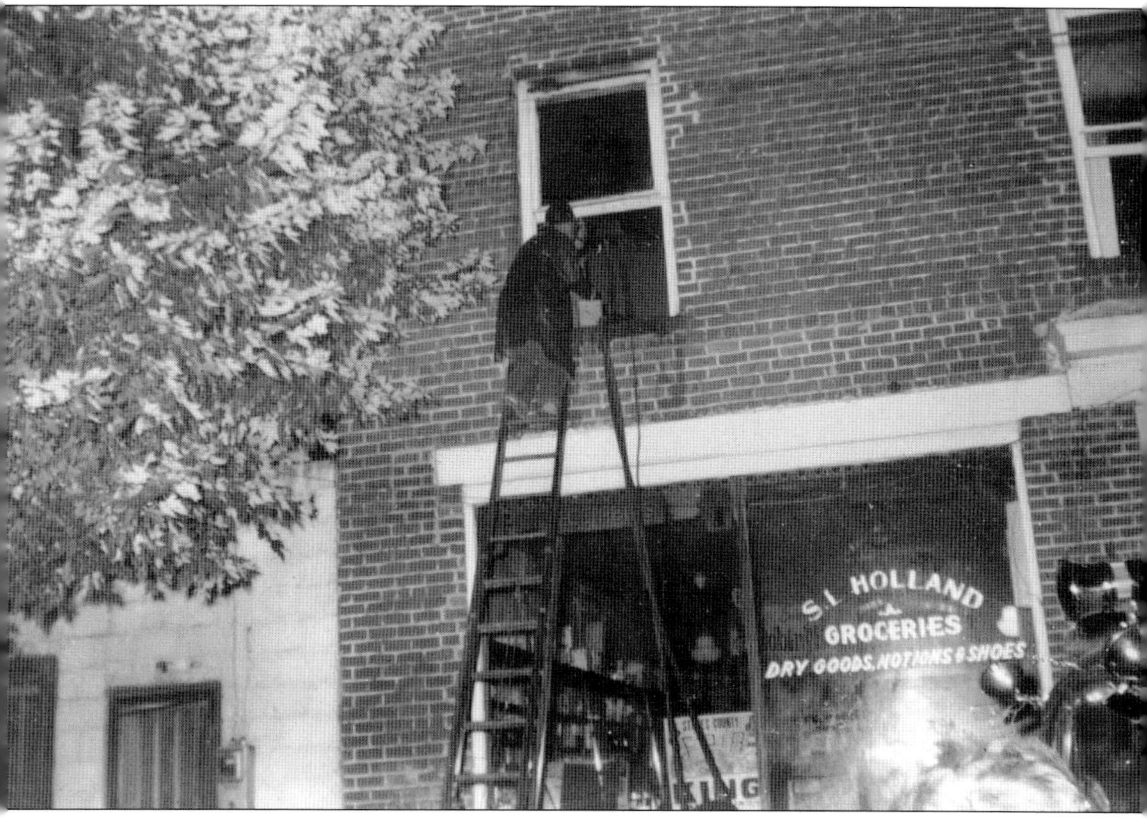

A firefighter from the Walkertown Volunteer Fire Department, Jesse Grubbs, is shown overhauling after a fire at the S.L. Holland grocery store. Grubbs owned Grubbs Garage in Walkertown. Many local business owners like Grubbs were volunteer firefighters. Many fires like this were caused by wood stoves used for heat during the winter. (Courtesy of FCFHS.)

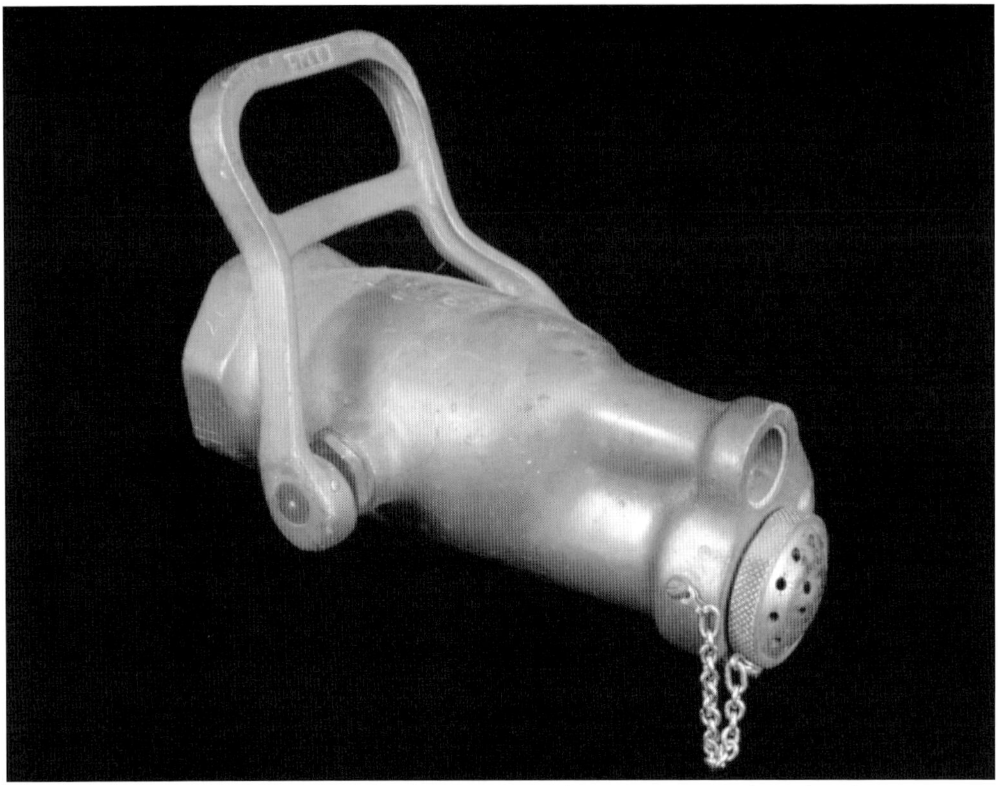

This is a Navy nozzle with a fog tip combination. The fog and straight stream nozzles were made according to US Navy specifications. All Forsyth County fire trucks had combination nozzles. (Courtesy of FCFHS.)

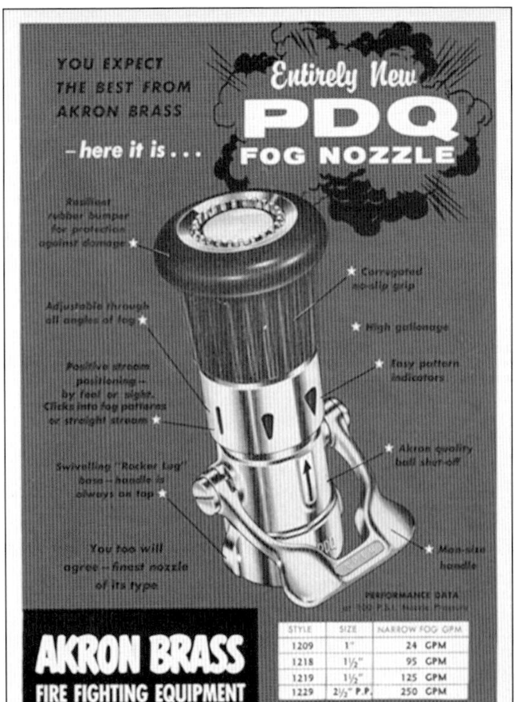

This is an early advertisement for the new Akron brass combination nozzle. The nozzle allowed for many different fog stream patterns. These nozzles were necessary to conserve water in rural areas. (Courtesy of FCFHS.)

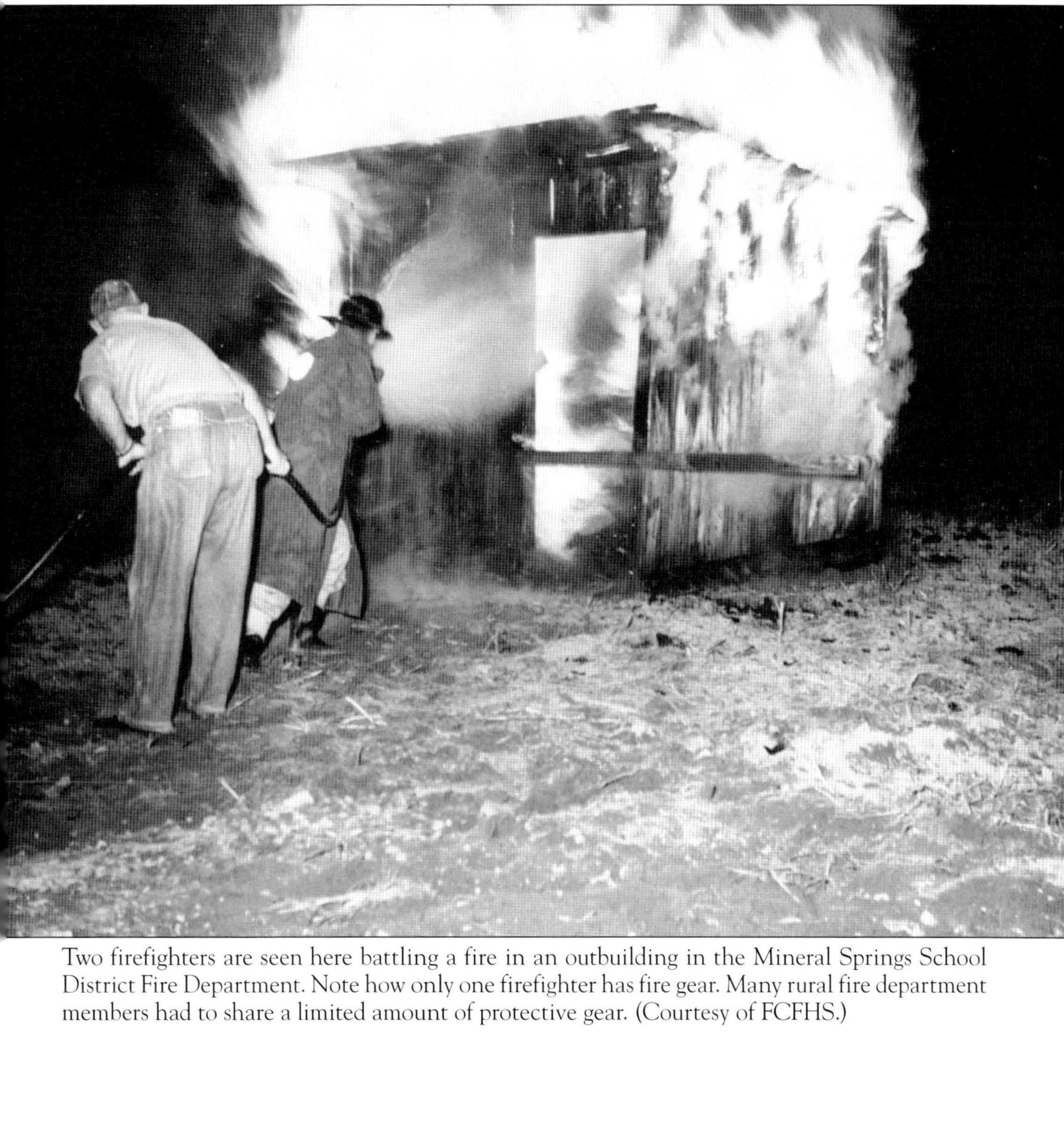

Two firefighters are seen here battling a fire in an outbuilding in the Mineral Springs School District Fire Department. Note how only one firefighter has fire gear. Many rural fire department members had to share a limited amount of protective gear. (Courtesy of FCFHS.)

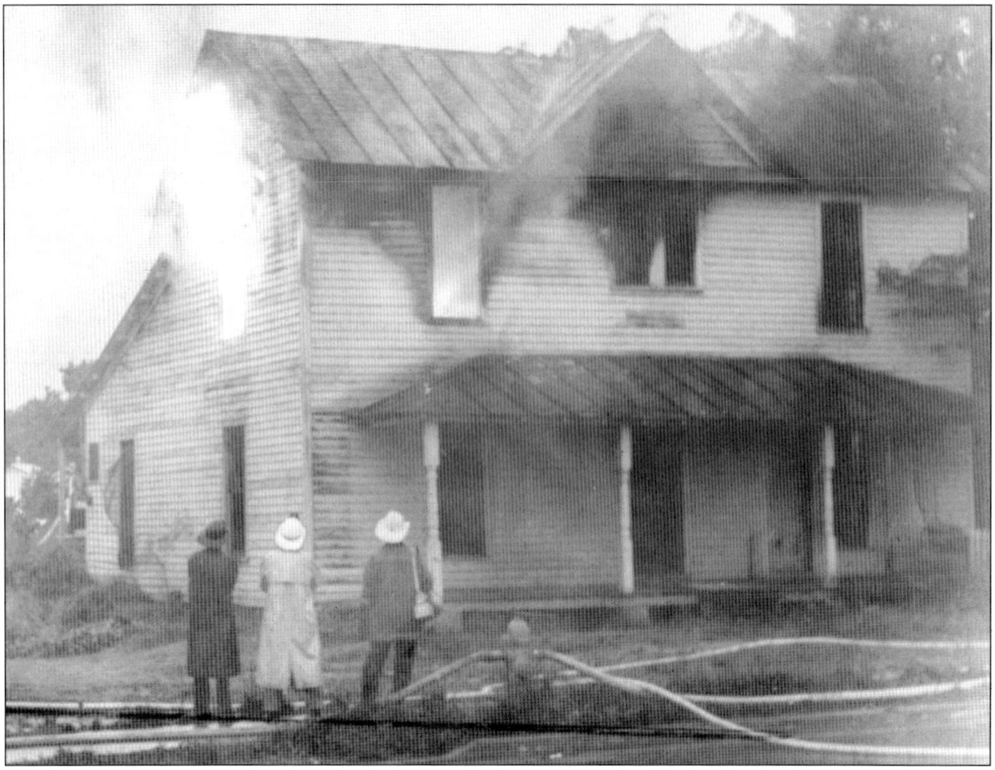

This house was used for a training burn where firefighters would learn about fire behavior. Obtaining a home to train in was a rare occurrence that brought many from the community together to learn. (Courtesy of FCFHS.)

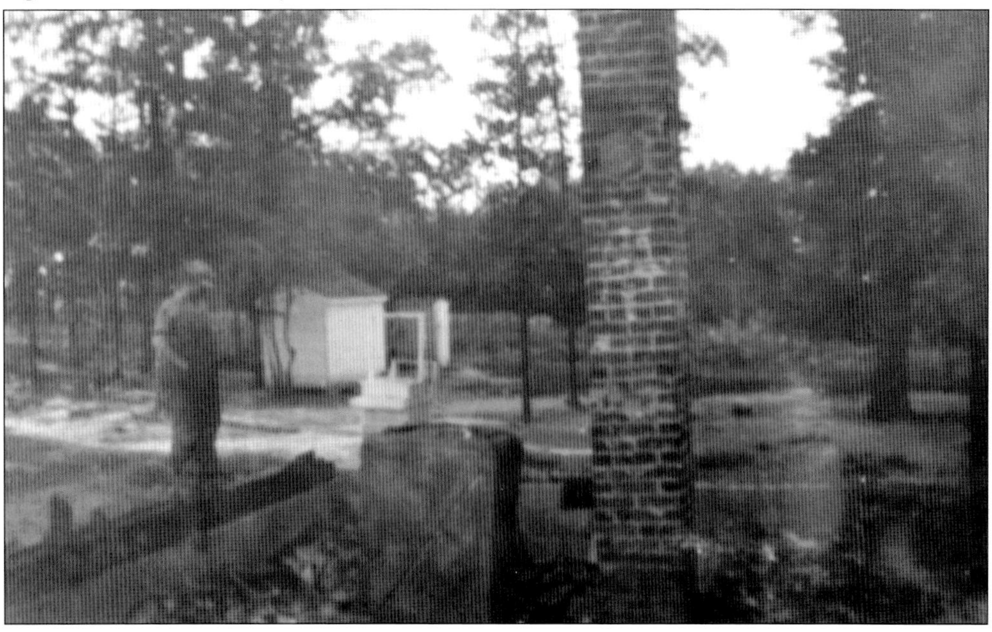

Shown here is another house fire that resulted in a total loss in rural Forsyth County. The need for fire equipment and training was growing with every fire incident. (Courtesy of FCFHS.)

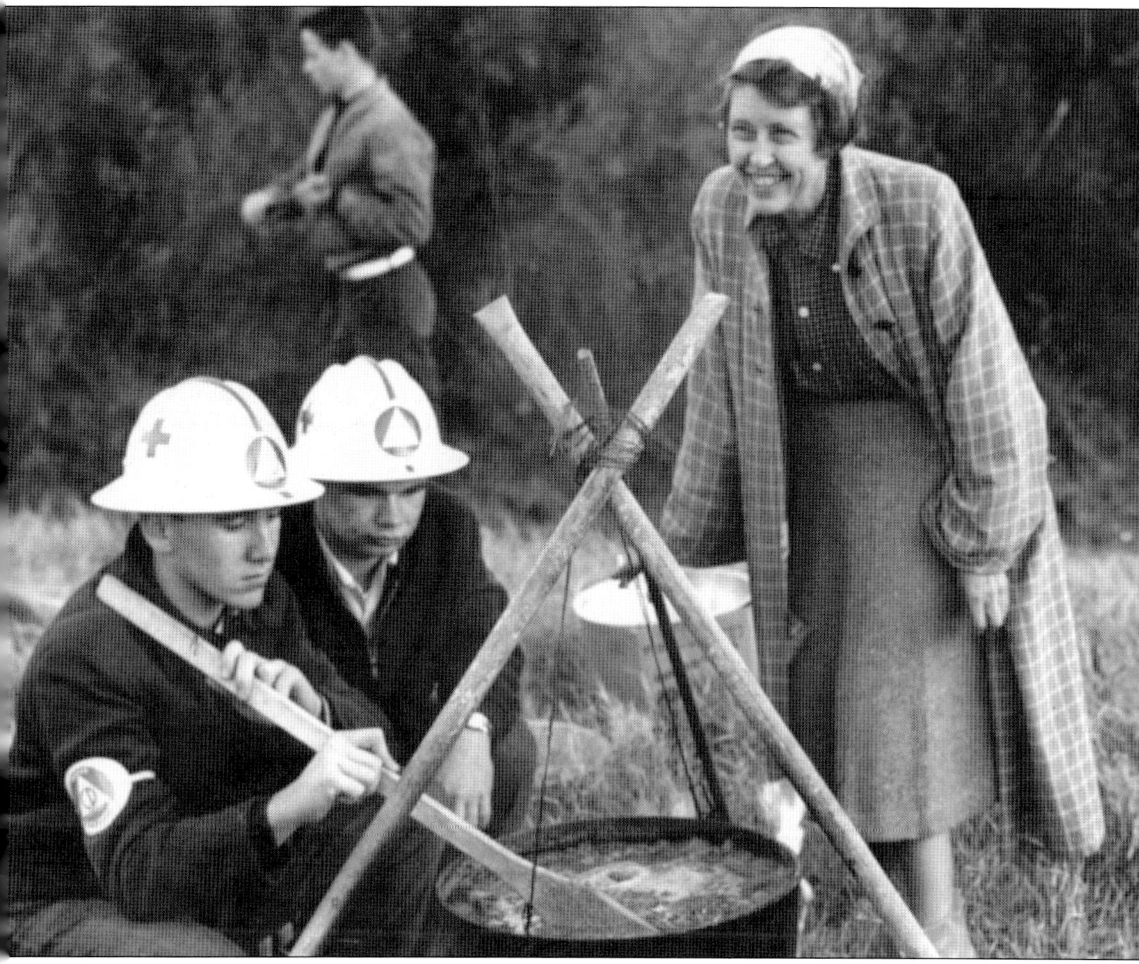

Explorer Scouts Jimmy Dorsett (left) and David Crockett conduct a test of the community's civil defense procedures with a simulated mock emergency while Mrs. Harold McKeithan of the Red Cross looks on. Civil defense activities in North Carolina were used to gain federal funding for infrastructure improvements. This was a tool to make the state a safer location for economic investments. Many of the early rural fire departments received grants from the Department of Civil Defense to purchase equipment and apparatus. (Courtesy FCPL.)

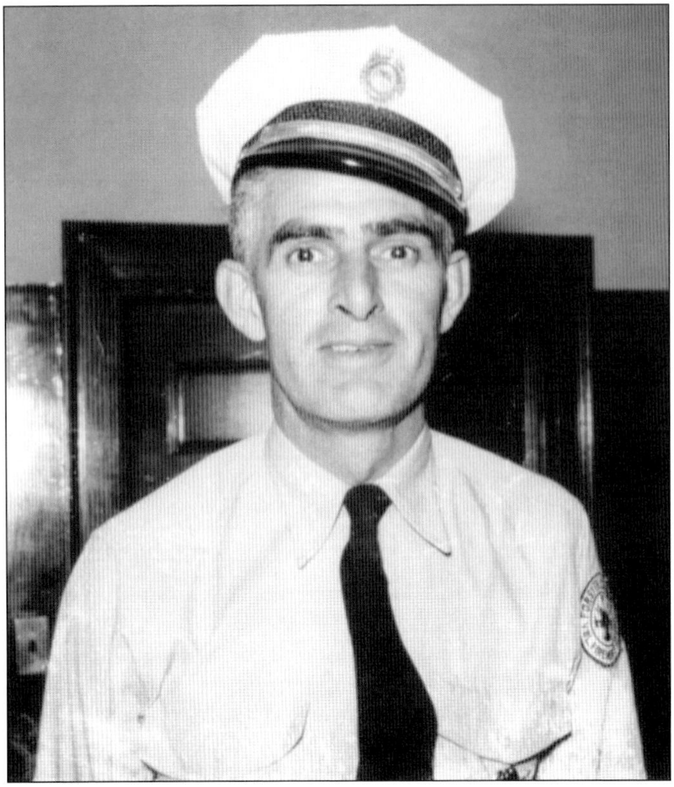

Pictured here is H.B. Self, who is credited with establishing the first local volunteer fire department. Self served as chairman of the Sedge Garden Volunteer Fire Department until he was elected chief in 1948. (Courtesy of FCFHS.)

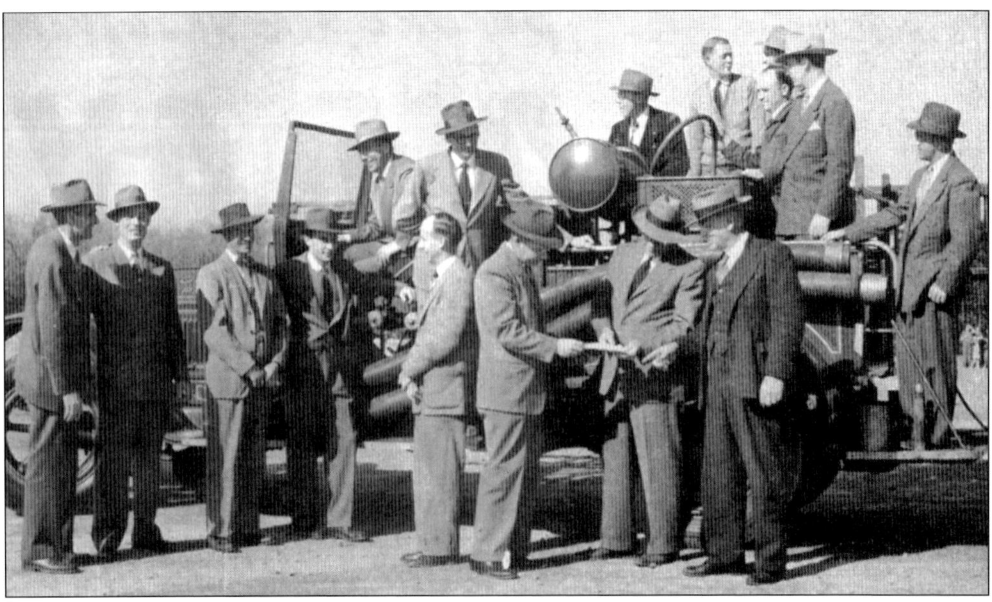

Sedge Garden Volunteer Fire Department members are standing by the fire truck they acquired from the Winston-Salem Fire Department. The truck was purchased through fundraisers and donations from the community. (Courtesy of FCFHS.)

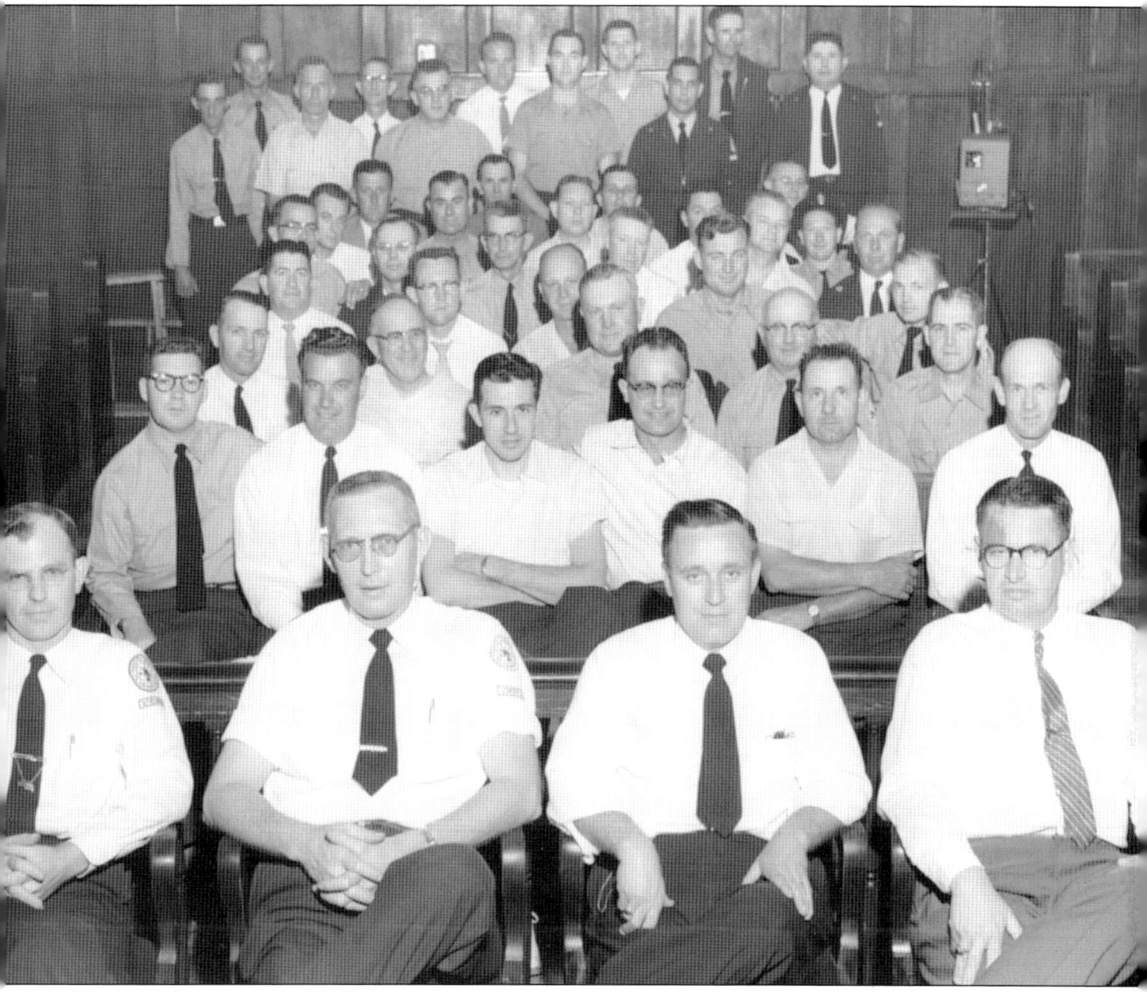

These men from various volunteer departments in Forsyth County gathered every month to discuss mutual aid training. Pictured from left to right at the Forsyth County Courthouse downtown in 1956 are (first row) J. Beeding, S. Needham, R. Bauguess, and L.E. Shingledecker; (second row) J. Johnson, H. Pardue, B. Saunders, R. Zimmerman, M. Garwood, and unidentified; (third row) D. Styers, E. Willard, and three unidentified; (fourth row) T. Davis, M. Fletcher, two unidentified, and D. Neal; (fifth row) P. Jennings, four unidentified, and M. Willson; (sixth row) G. Hildreth, A. Dillon, G. Ray, D. Conrad, and unidentified; (seventh row) G. Sisk, W. Tatum, C. Webster, and unidentified; (eighth row) J. Hampton, R. Long, H. Beeson, B. Moser, J.E. Mecum, and T. Trout; (ninth row) L. Sapp, three unidentified, and R. Bowles. (Courtesy of FCPL.)

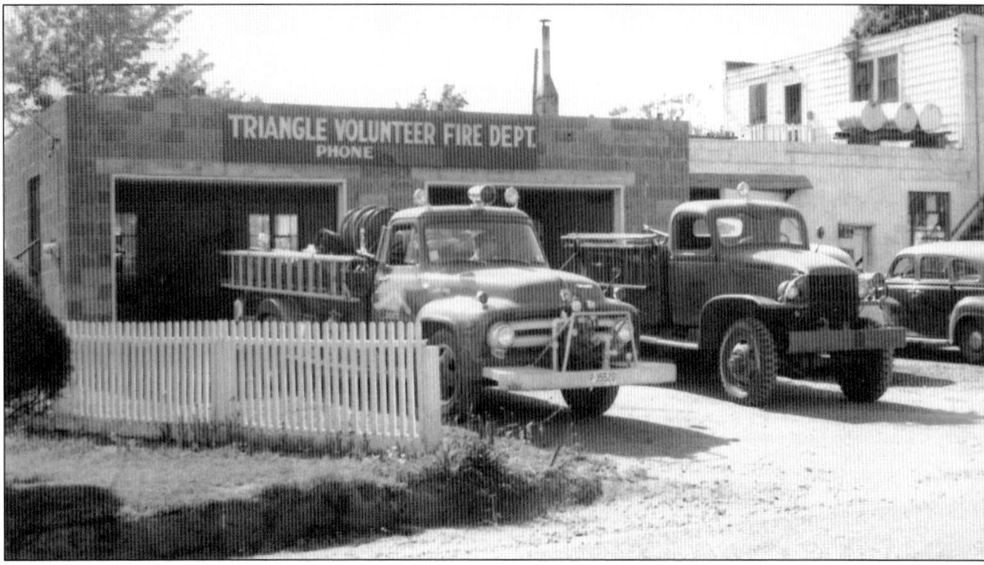

Pictured is the Triangle Volunteer Fire Department (formerly known as the Sedge Garden Volunteer Fire Department). This department served not only the Triangle/Sedge Garden fire district, but also Union Cross and Beeson Cross Roads before those areas established fire protection. (Courtesy of FCFHS.)

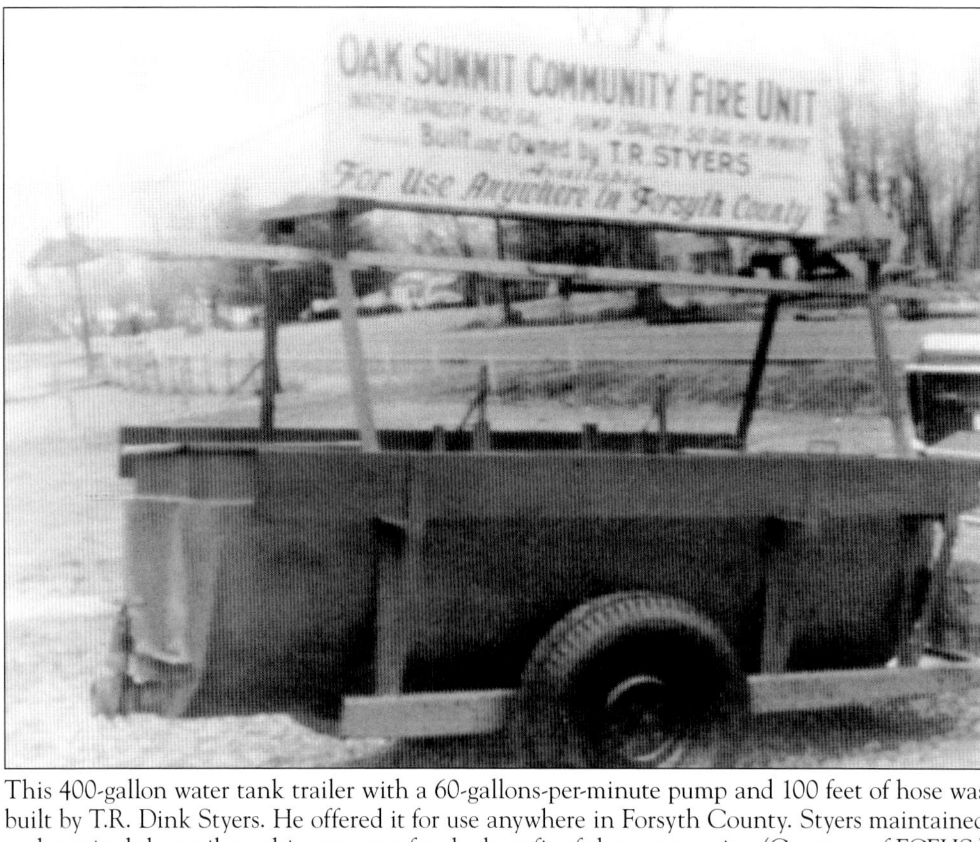

This 400-gallon water tank trailer with a 60-gallons-per-minute pump and 100 feet of hose was built by T.R. Dink Styers. He offered it for use anywhere in Forsyth County. Styers maintained and repaired the trailer at his own cost for the benefit of the community. (Courtesy of FCFHS.)

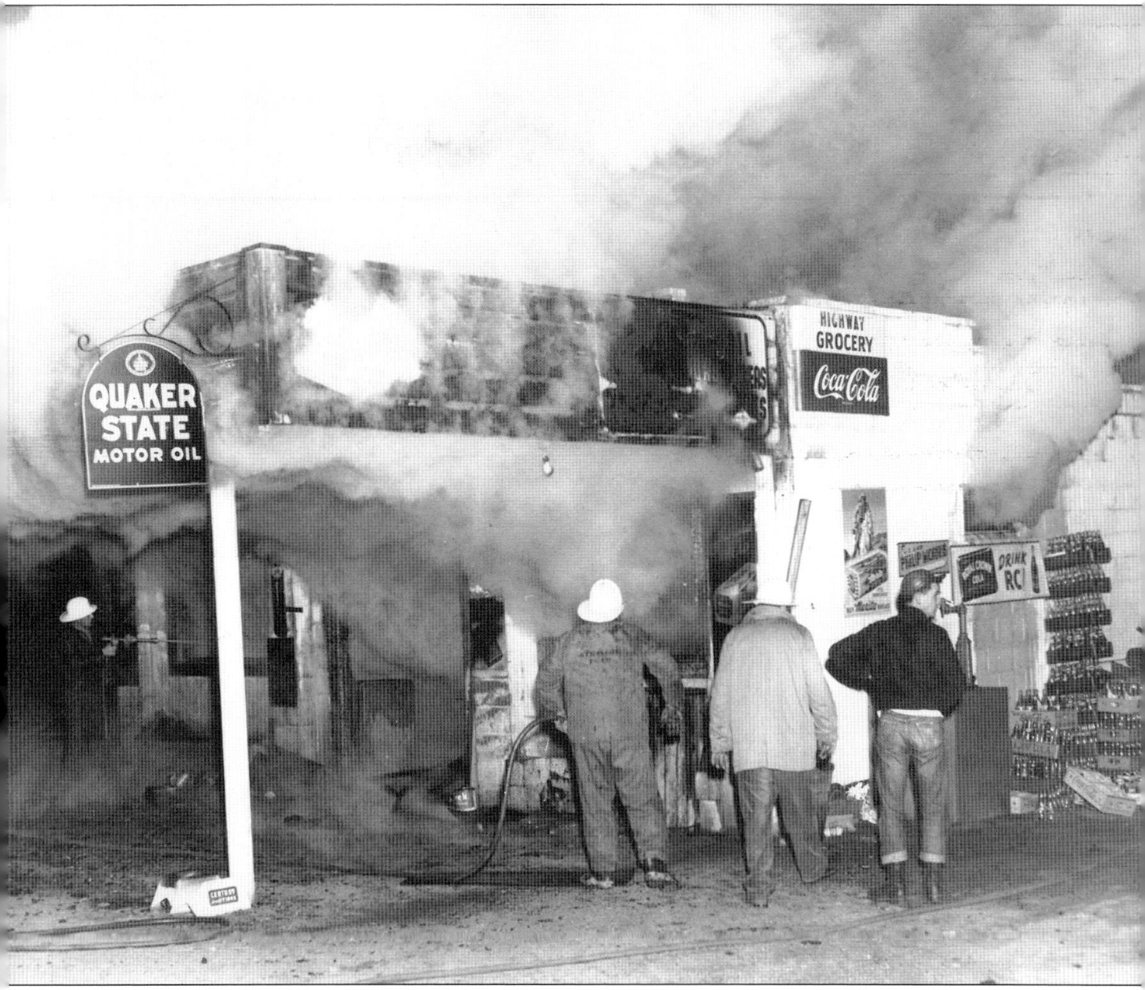

This large fire broke out at a service station in Walkertown, causing extensive damage. Local business owners had a vested interest in establishing better and more effective means of fire protection. (Courtesy of FCFHS.)

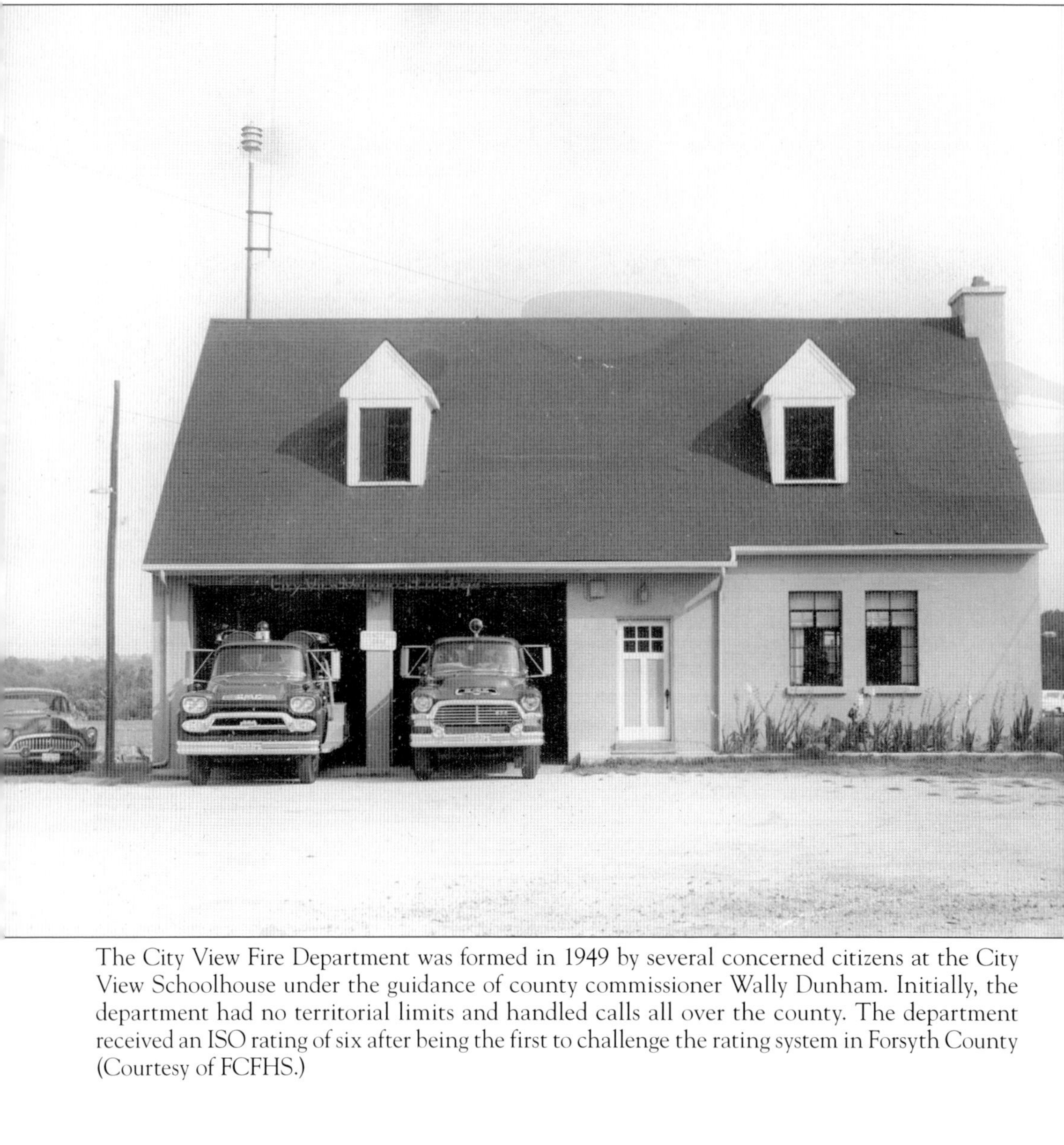

The City View Fire Department was formed in 1949 by several concerned citizens at the City View Schoolhouse under the guidance of county commissioner Wally Dunham. Initially, the department had no territorial limits and handled calls all over the county. The department received an ISO rating of six after being the first to challenge the rating system in Forsyth County (Courtesy of FCFHS.)

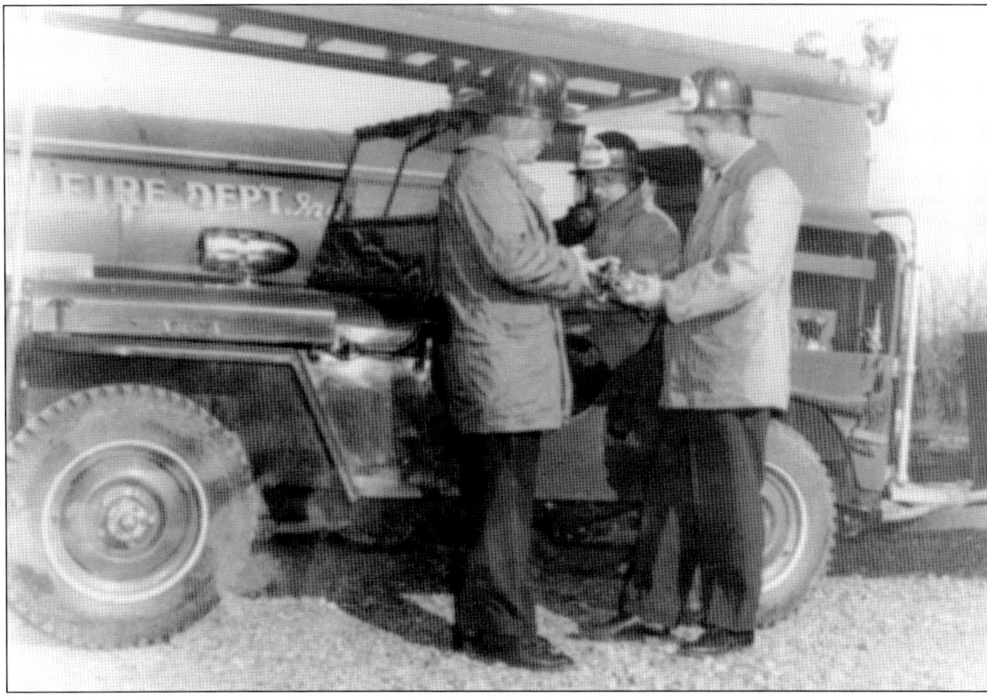

City View Volunteer Fire Department personnel are looking over new fire equipment. Equipment like the Navy fog nozzle was simple to operate and train on. From left to right are Wally Dunham, Walt Idol, and Chief Charles Macum. (Courtesy of FCFHS.)

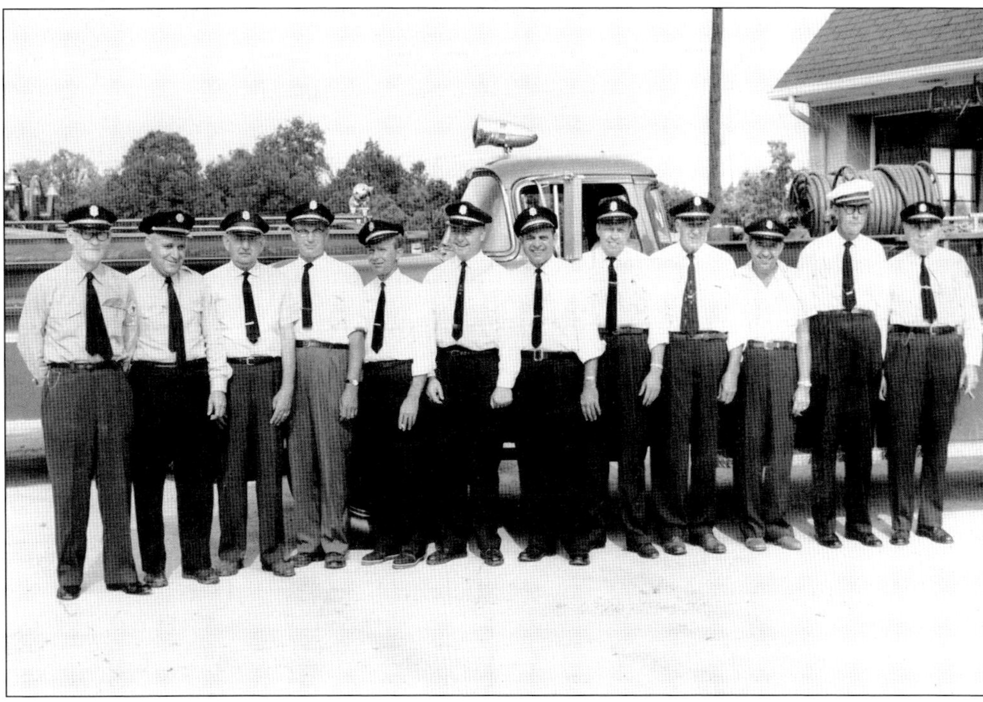

Pictured are several founding members of the City View Volunteer Fire Department. Chief H.G. Peddycord is second from the left. (Courtesy of FCFHS.)

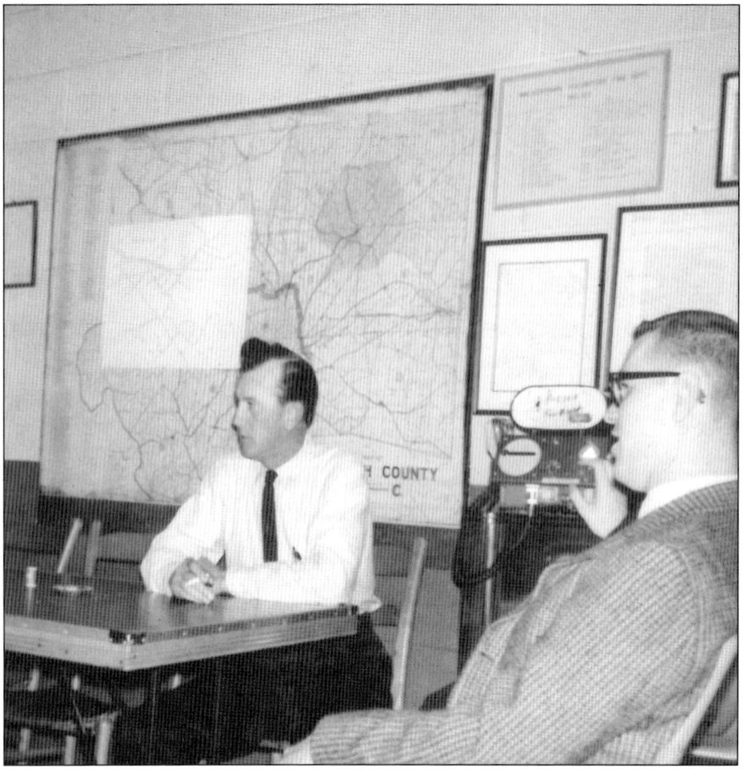

Pictured is Herbert Oakley of Mineral Springs School District Fire Department attending a rural fire protection meeting in 1950. These meetings were catalysts in the development of training programs for volunteer firefighters. (Courtesy of FCFHS.)

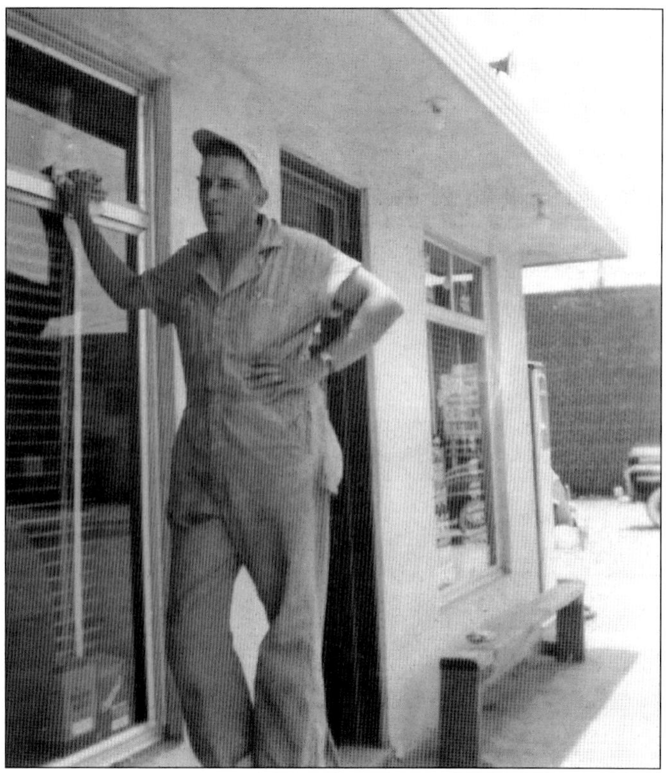

Clyde Webster is shown wearing his service station uniform. Webster was a volunteer firefighter and went on to be the first fire chief of the Walkertown Volunteer Fire Department. (Courtesy of FCFHS.)

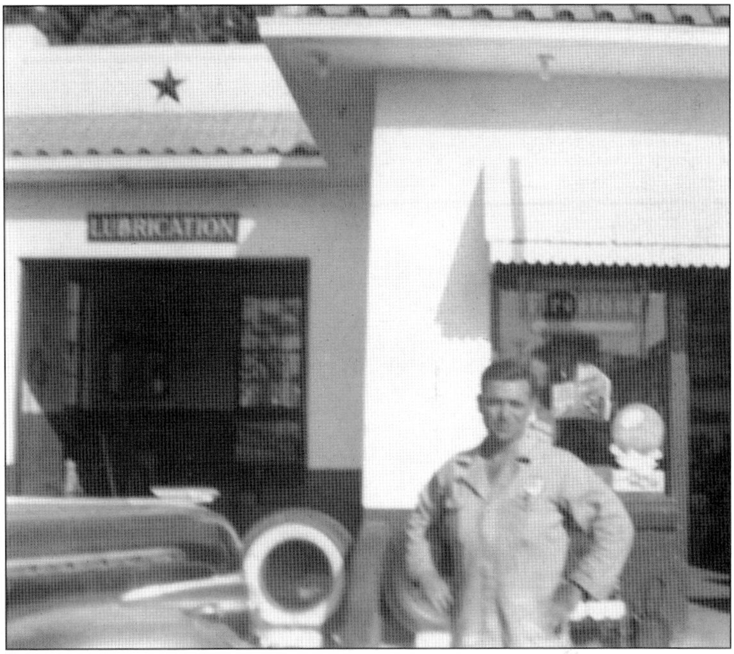

Rick Waggoner is shown in front of his service station on Stadium Drive. Waggoner also owned Rick's Restaurant in Walkertown. Like Clyde Webster, Waggoner was a community member, business owner, and volunteer firefighter willing to drop whatever he was doing and respond to fire calls in the area. (Courtesy of FCFHS.)

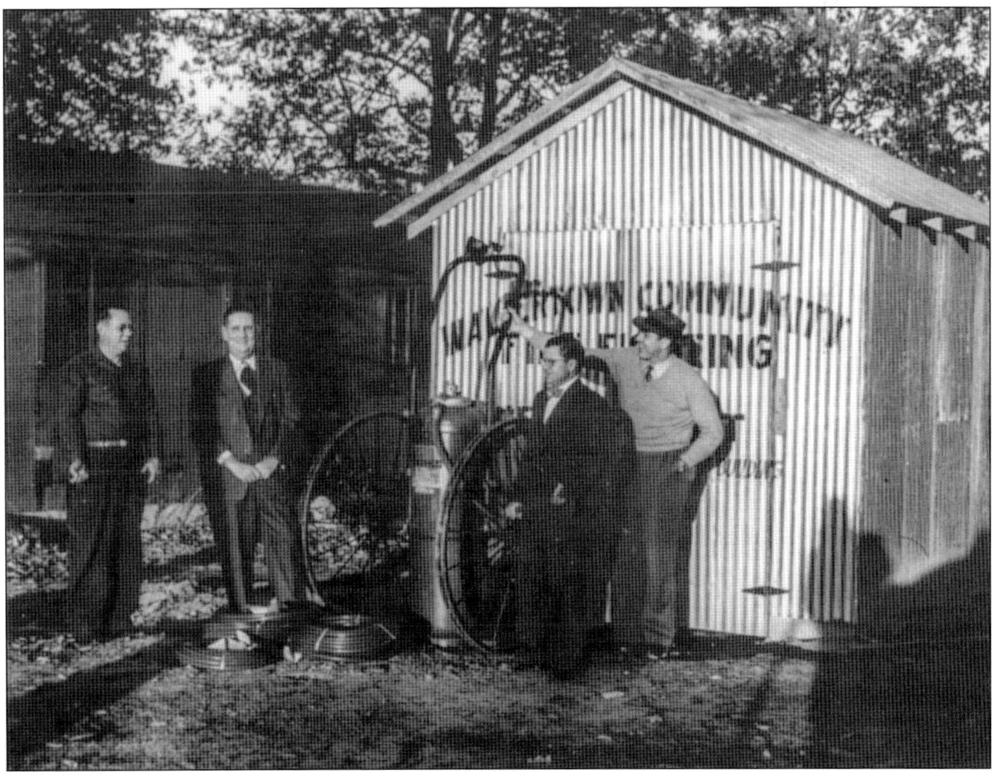

The Walkertown Civic Club recommended to the town council that funds be acquired to establish a fire department on February 27, 1953. Pictured is the Walkertown community firefighting equipment storage building with volunteer members (from left to right) James Gant, Phil Booe, George Leight, and Robert Hutcherson with the department's fire equipment. (Courtesy of FCFHS.)

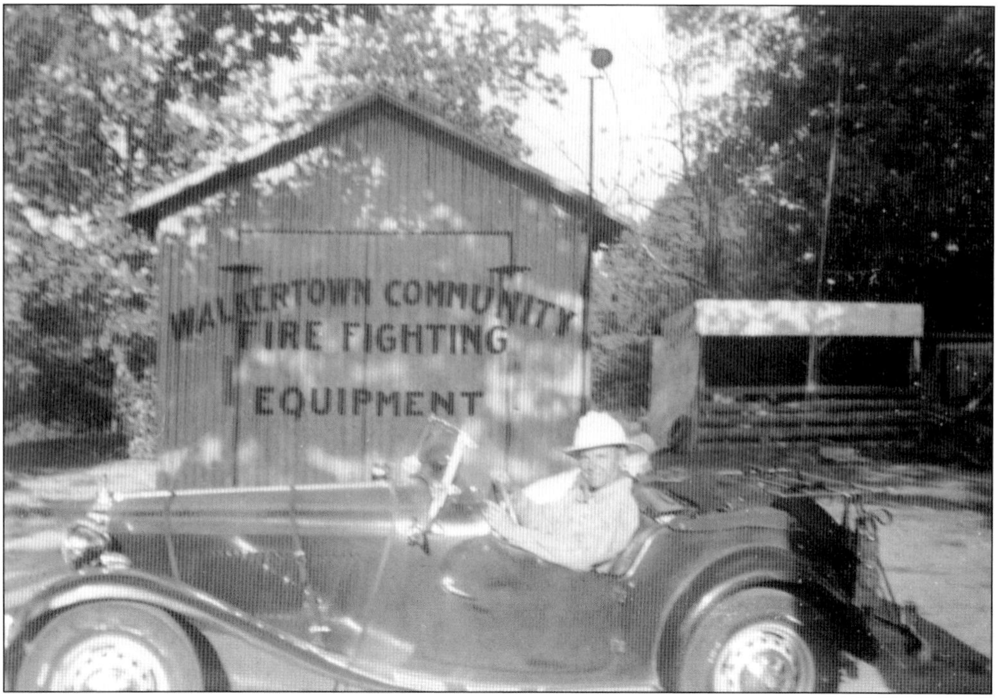

Rick Waggoner is seen here in front of the original Walkertown Volunteer Fire Department's equipment building. (Courtesy of FCFHS.)

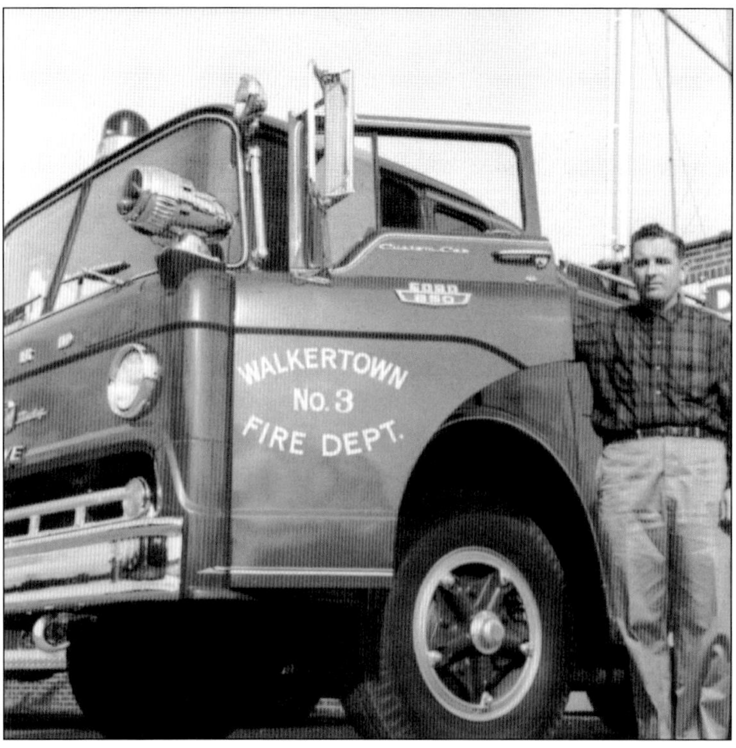

Chief Clyde Webster is shown standing next to a newly acquired 1963 Ford apparatus with Howe firefighting equipment. (Courtesy of FCFHS.)

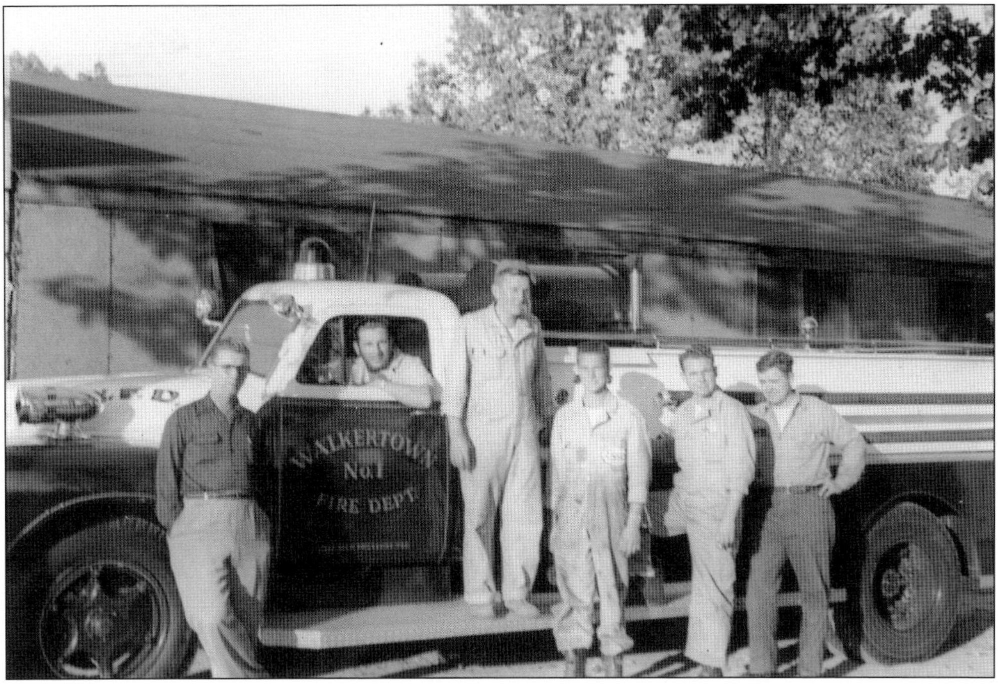

Members of the Walkertown Volunteer Fire Department are seen here with the department's 1953 GMC fire truck. (Courtesy of FCFHS.)

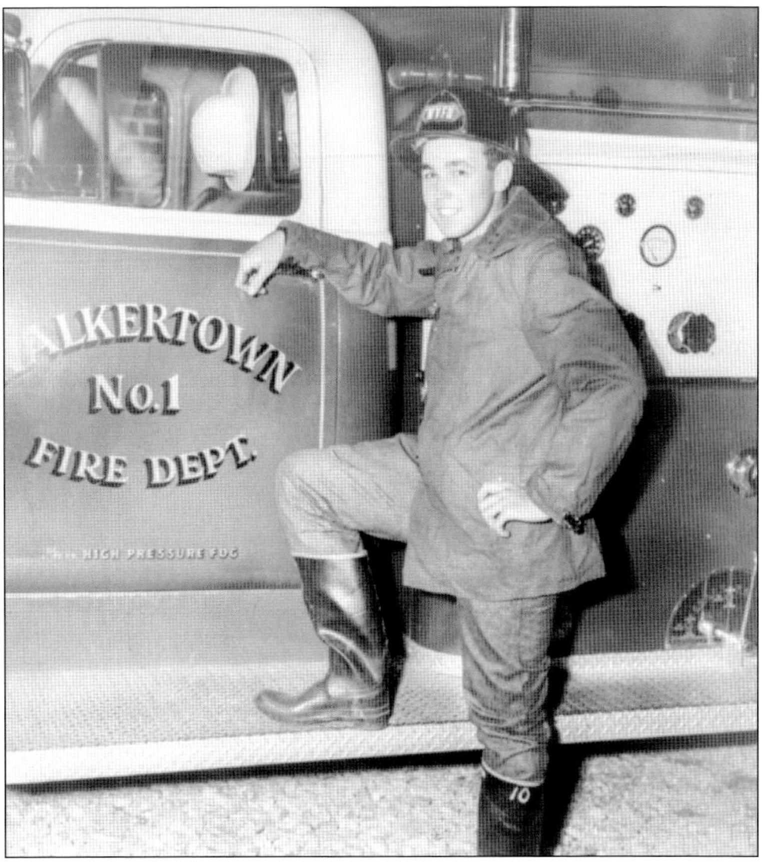

Volunteer member Russell Gray Fulp is proudly standing next to the Walkertown Volunteer Fire Department's 1953 GMC fire truck with John Bean equipment. Fulp was also a member of the Walkertown Civic Club. (Courtesy of FCFHS.)

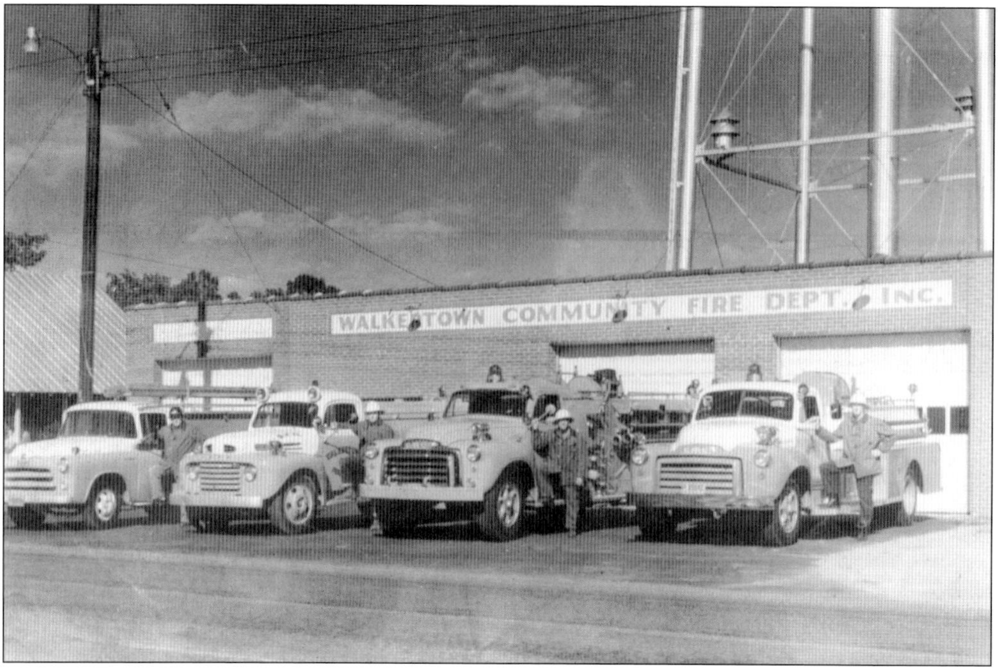

In 1953, the Walkertown Volunteer Fire Department obtained a new building in central Walkertown. The department also purchased various fire apparatus. (Courtesy of FCFHS.)

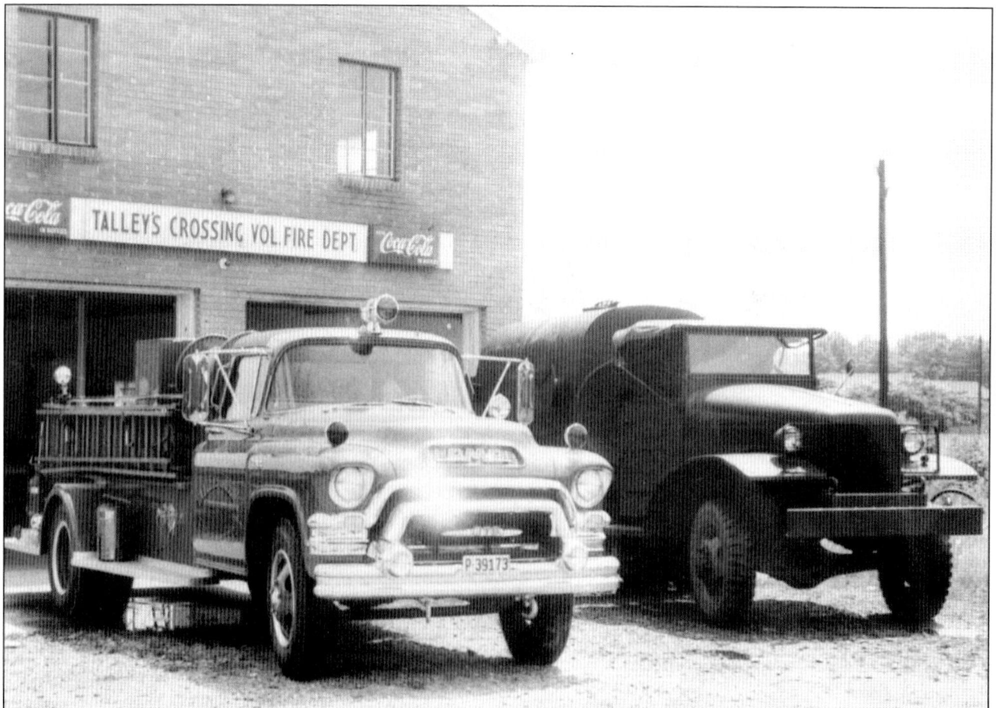

The Talley's Crossing Fire Department building was purchased with community member donations and pledges in 1954. The build was motivated by a large wildfire that was extinguished by the local residents using only shovels and rakes. (Courtesy of FCFHS.)

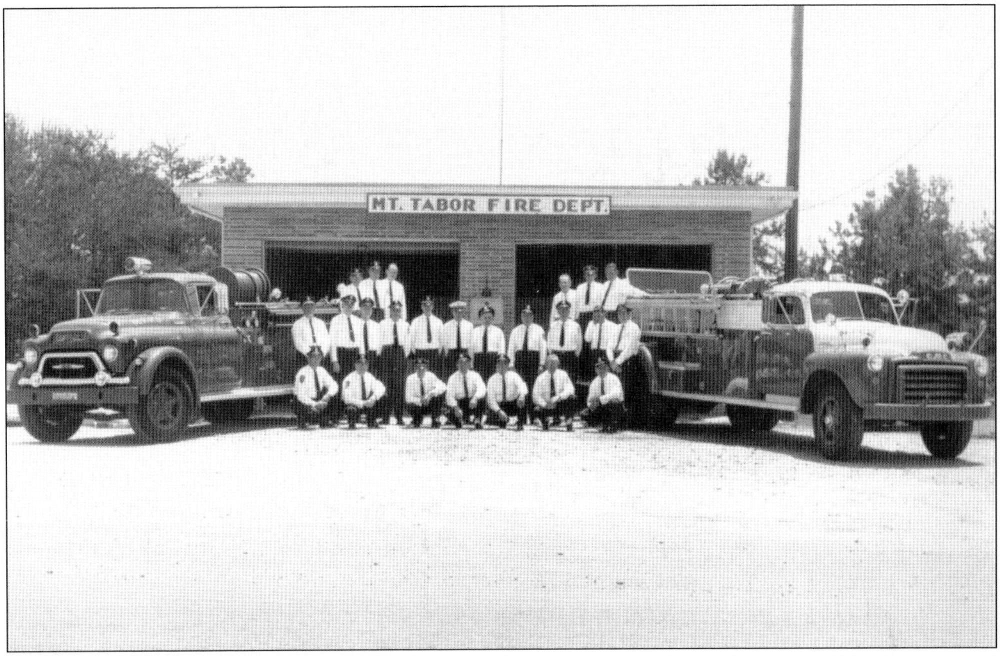

The Mount Tabor Fire Department building and apparatus are shown here in 1954. The original fire apparatus was purchased from donations from the community. The department building was constructed on land leased from Carl Sapp, a member of the Mount Tabor Civic Club. (Courtesy of FCFHS.)

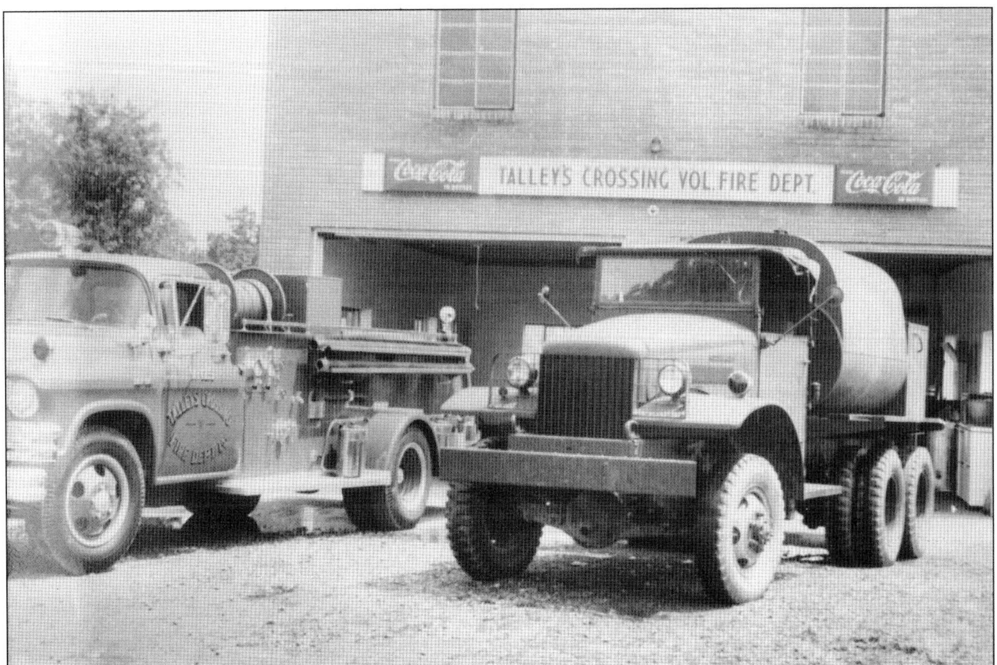

Talley's Crossing Fire Department purchased this 1954 International 1,200-gallon military tanker vehicle in 1957. The 1,200-gallon tanker was a great asset due to fire hydrants not being available in the rural districts. (Courtesy of FCFHS.)

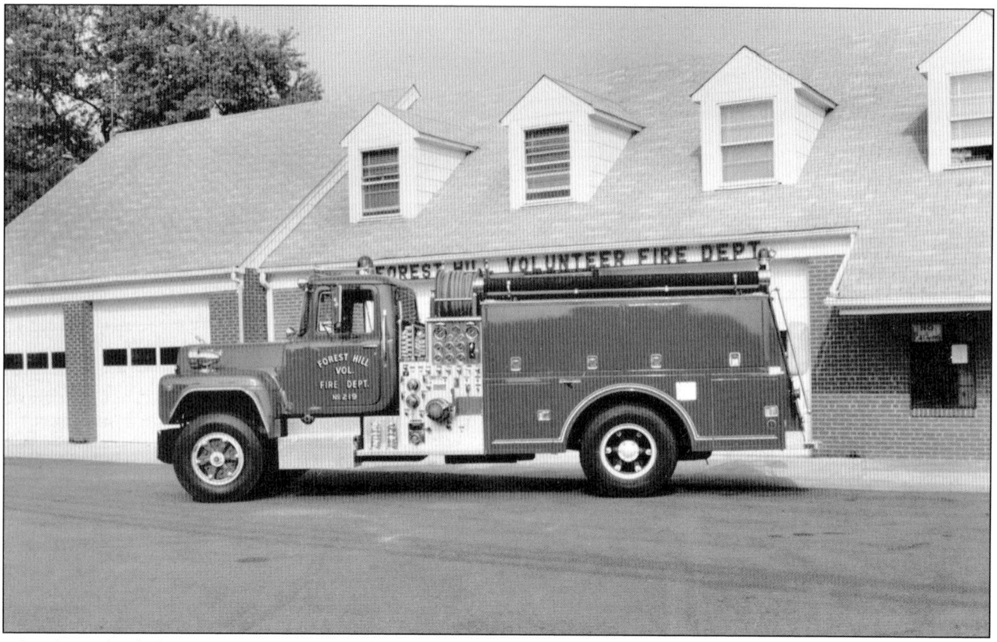

The board of directors of the Forest Hill Civic Club met in 1951 and discussed the formation of a fire department. In 1952, representatives of local businesses as well as residents and county commissioner Wally Dunham shared ideas that would later develop into the Forest Hill Volunteer Fire Department. The department building was constructed on North Cherry Street in 1954. (Courtesy of FCFHS.)

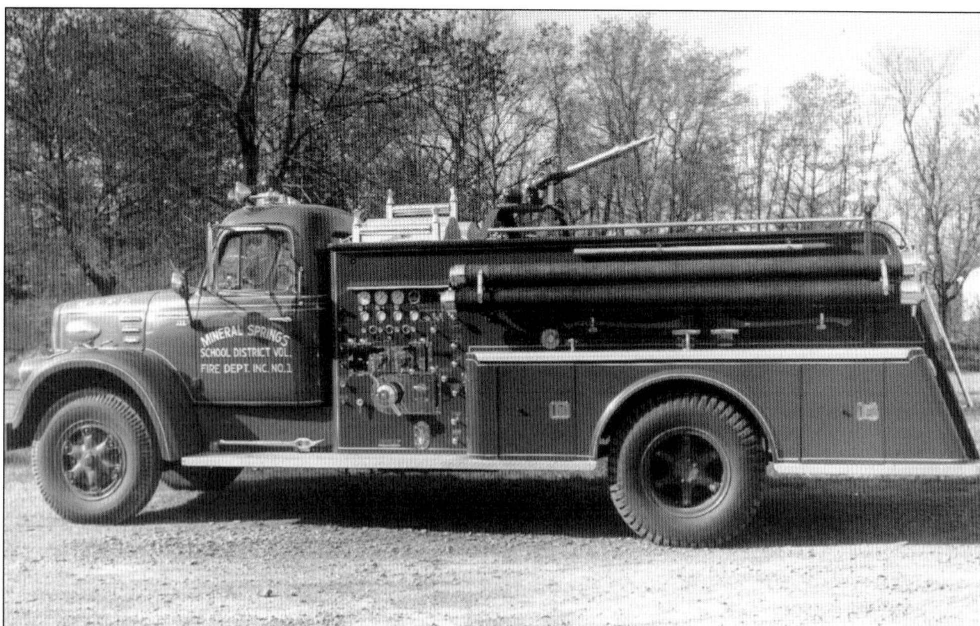

The Mineral Springs School District Fire Department apparatus was located at a lot purchased by members of the Mineral Springs Civic Club, the White Rock Compliment Club, and the Ogburn Station Fire Protection Association. The department building was constructed later in 1953. (Courtesy of FCFHS.)

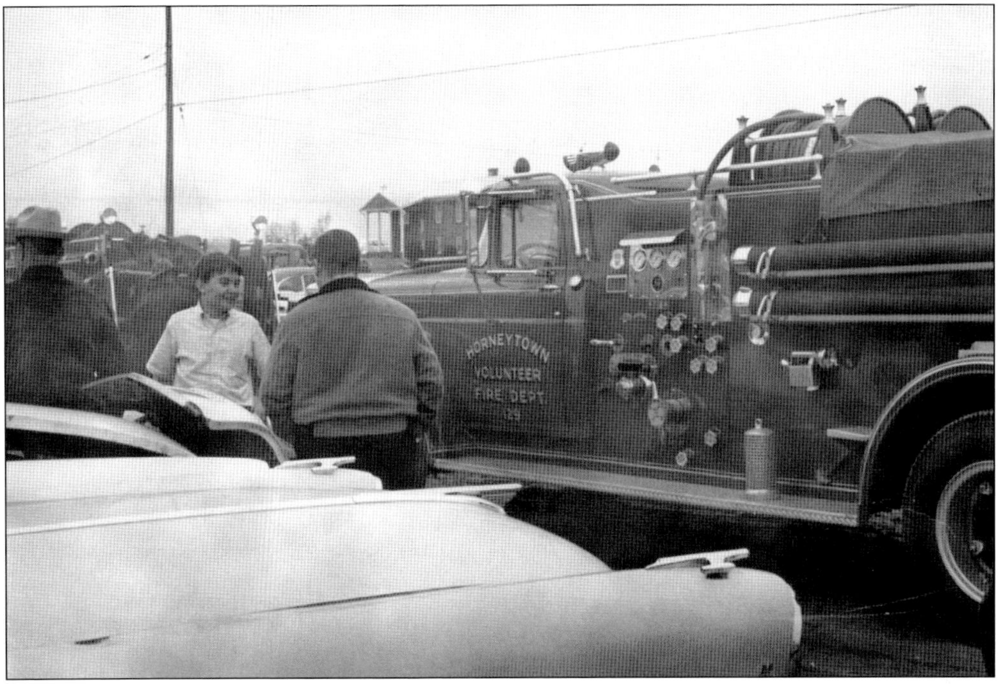

The 1959 Ford Howe 500-gallons-per-minute pumper of the Horneytown Volunteer Fire Department is seen above at an incident in a rural part of the county. This pumper was delivered to the department just three days before it responded to its first fire. Below, the Union Cross Volunteer Fire Department is shown assisting the Horneytown Volunteer Fire Department on a working fire event in 1959. This event demonstrated the importance of mutual aid training with neighboring fire departments. (Both, courtesy of FCFHS.)

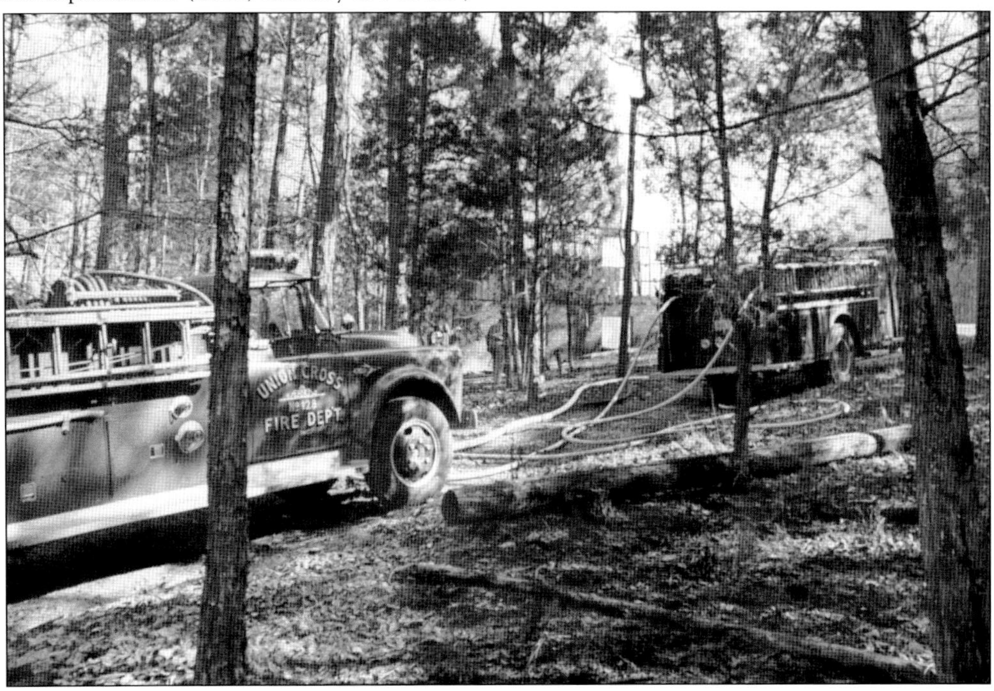

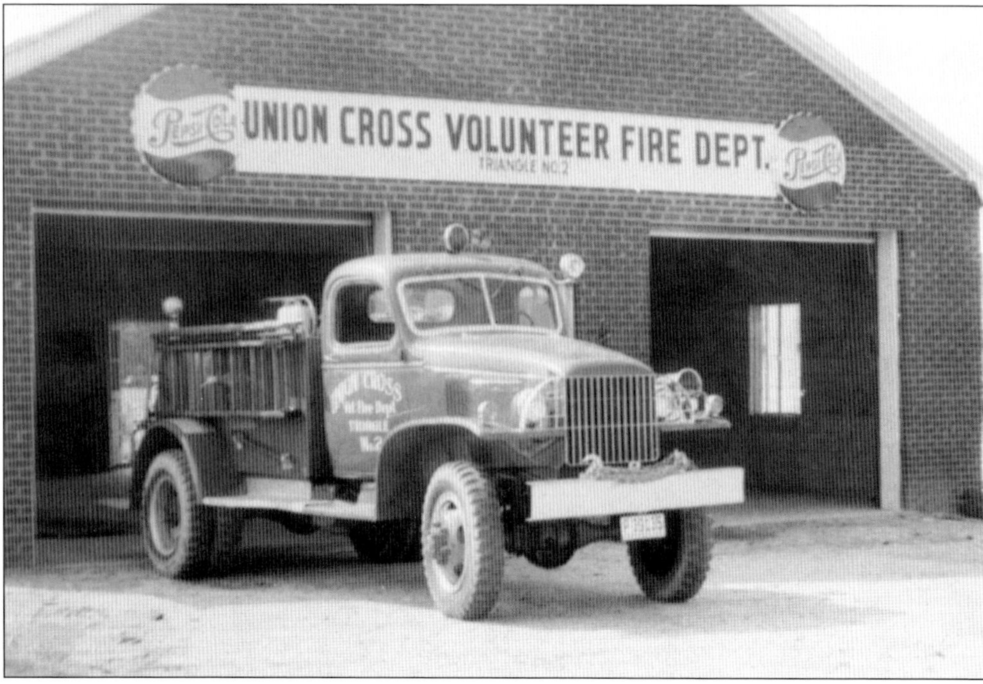

The Union Cross Fire Department building is shown here in 1958. This department began as an auxiliary unit of the Triangle Volunteer Fire Department in 1957. The 1957 eight-hundred-gallon Dodge truck pictured here was housed on Union Cross Road in a building owned by the Union Cross Moravian Church. (Courtesy of FCFHS.)

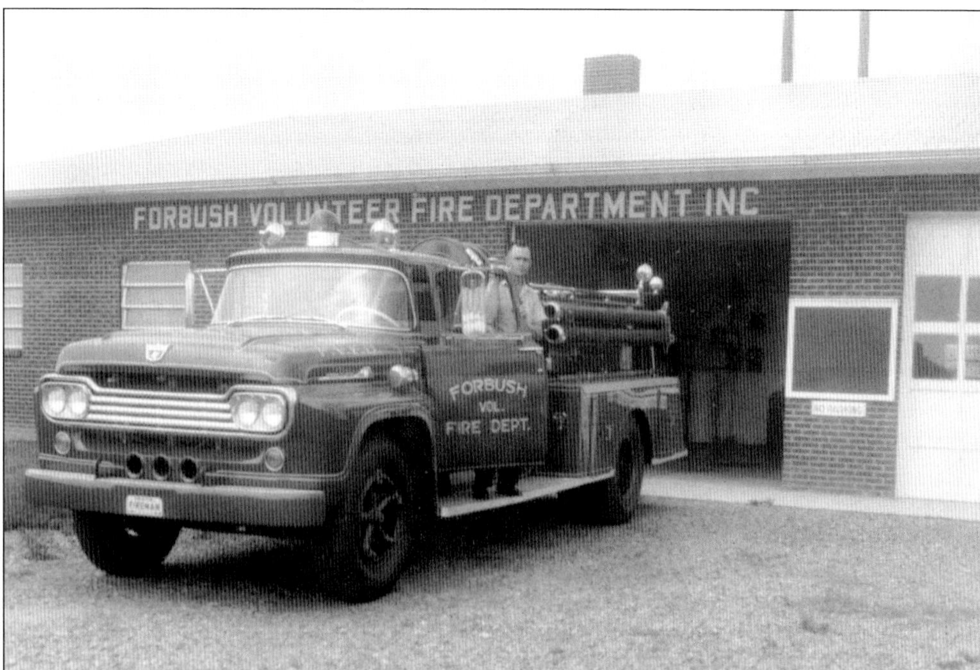

The Forbush Volunteer Fire Department building was constructed in 1960. Assistant Chief Fred Dinkins is pictured here with the department's first engine. (Courtesy of FCFHS.)

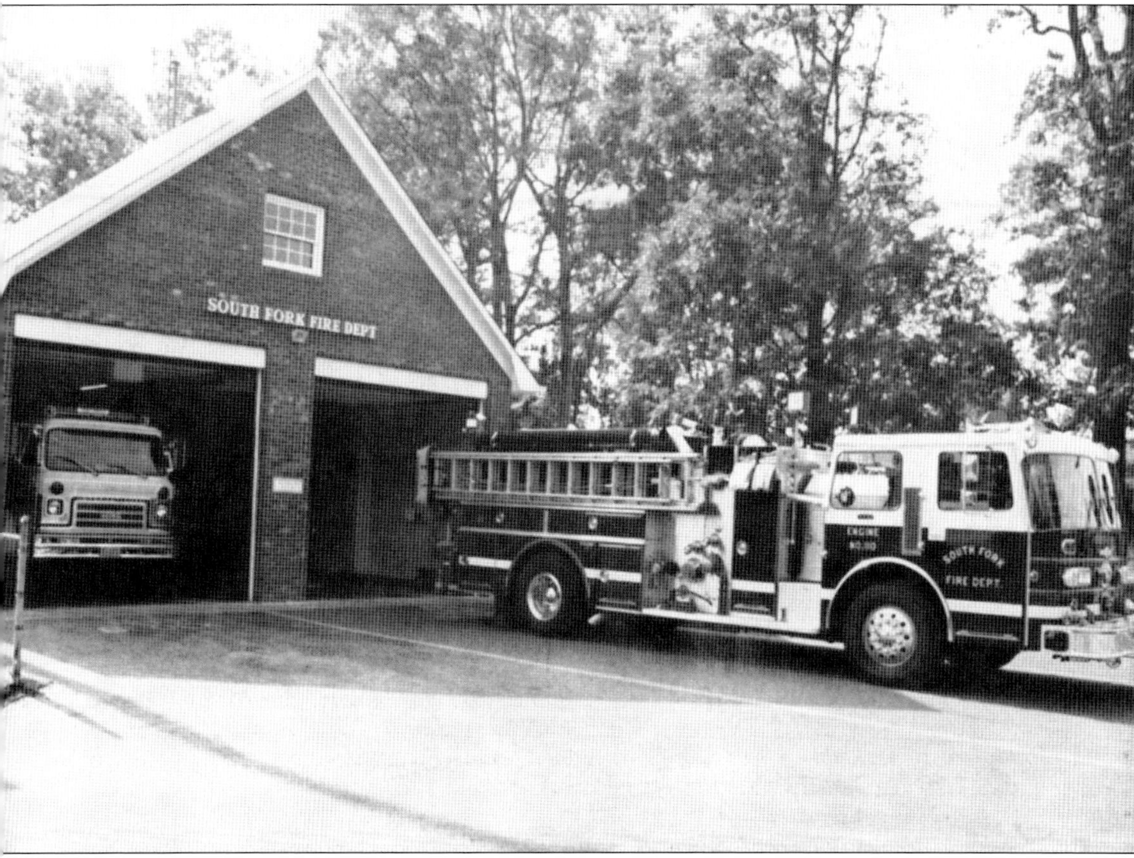

The South Fork Volunteer Fire Department was founded in 1951. It was one of the first rural departments to receive a charter. In 1990, South Fork merged with the Clemmons Fire Department. (Courtesy of FCFHS.)

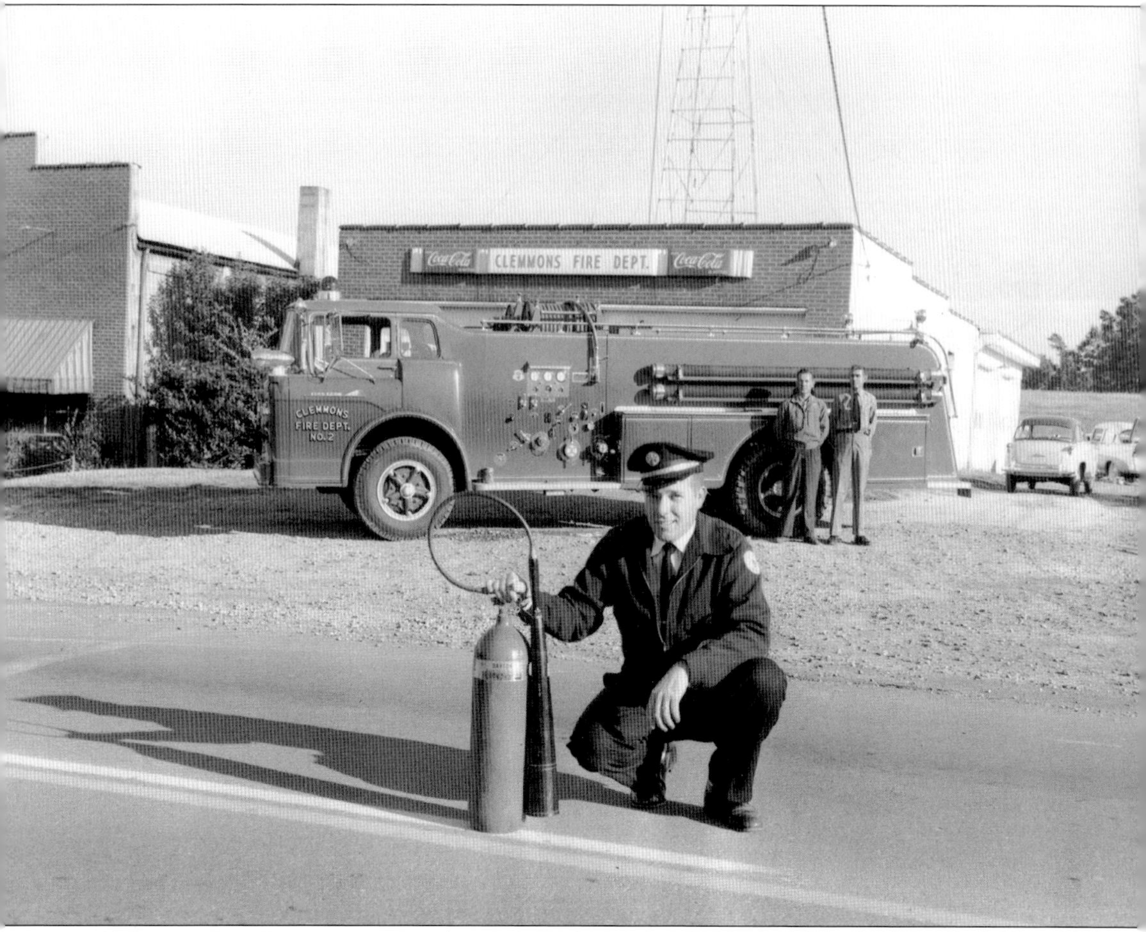

Clemmons Fire Department member "Big" Jack Higgins displays a fire extinguisher in front of the station in 1960. Higgins was a local business owner and served as president of the Clemmons Civic Club. (Courtesy of FCPL.)

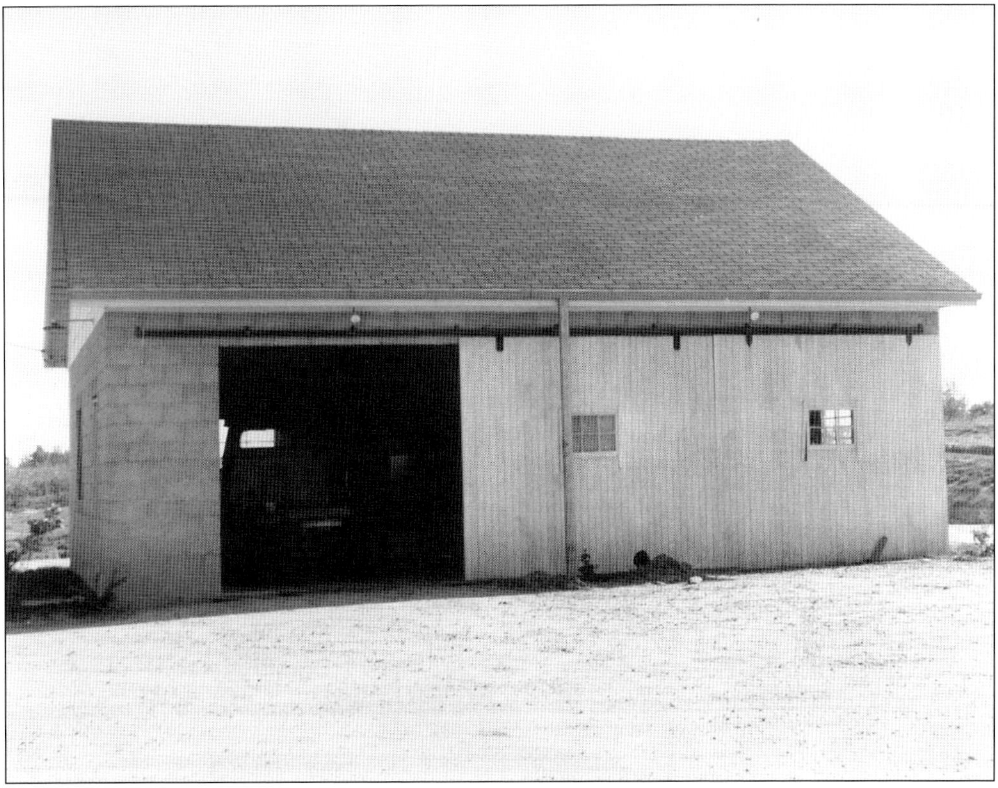

The Beeson Cross Roads Volunteer Fire Department building is shown here in 1954. The department signs were posted on the building later. The signs were often sponsored by local businesses bearing their name or logo. (Courtesy of FCFHS.)

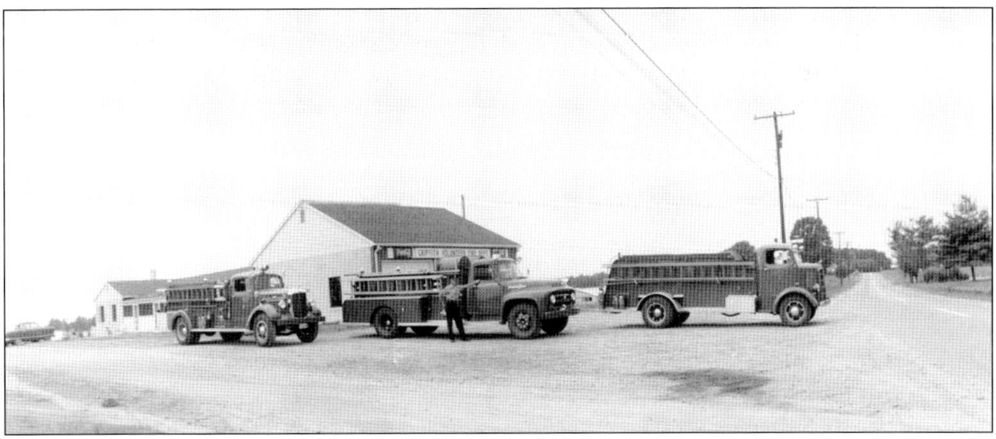

The Griffith Volunteer Fire Department is seen here in August 1957. The building was on West Clemmonsville Road. This department was organized in 1952 after a meeting at the Griffith schoolhouse where Lewisville fire chief Buck Reynolds and county commissioner Wally Dunahm spoke about the process needed to form a department. From this meeting, the government body was formed and included such members as G.B. Tuttle, W.F. Kern, Glenn Ray, Clinton Lee, Crouse Jones, Raymond Hill, and Albert Berrier. Charles Fox was named fire chief of the department. (Courtesy of FCPL.)

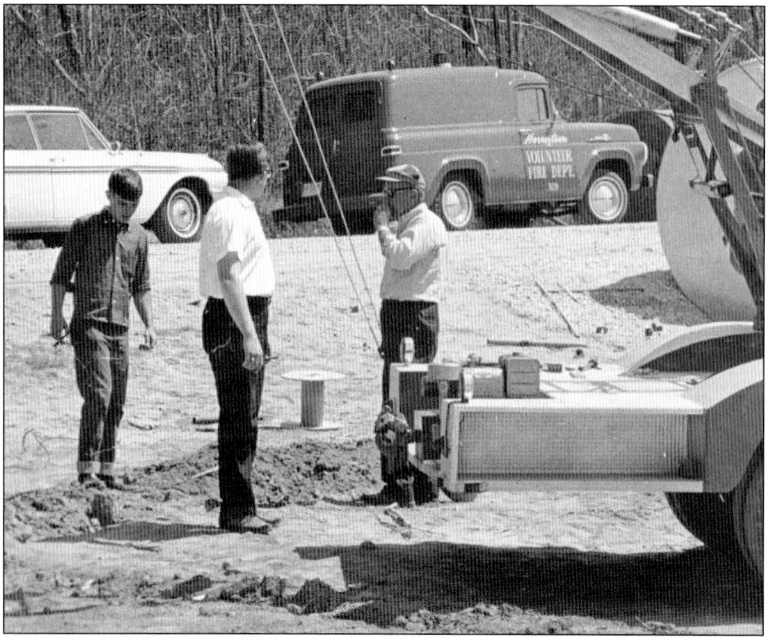

Horneytown Volunteer Fire Department members are pictured here on the scene of an event. The department was started in 1958. The fire district included Davidson and Guilford Counties in addition to Forsyth County. (Courtesy of FCFHS.)

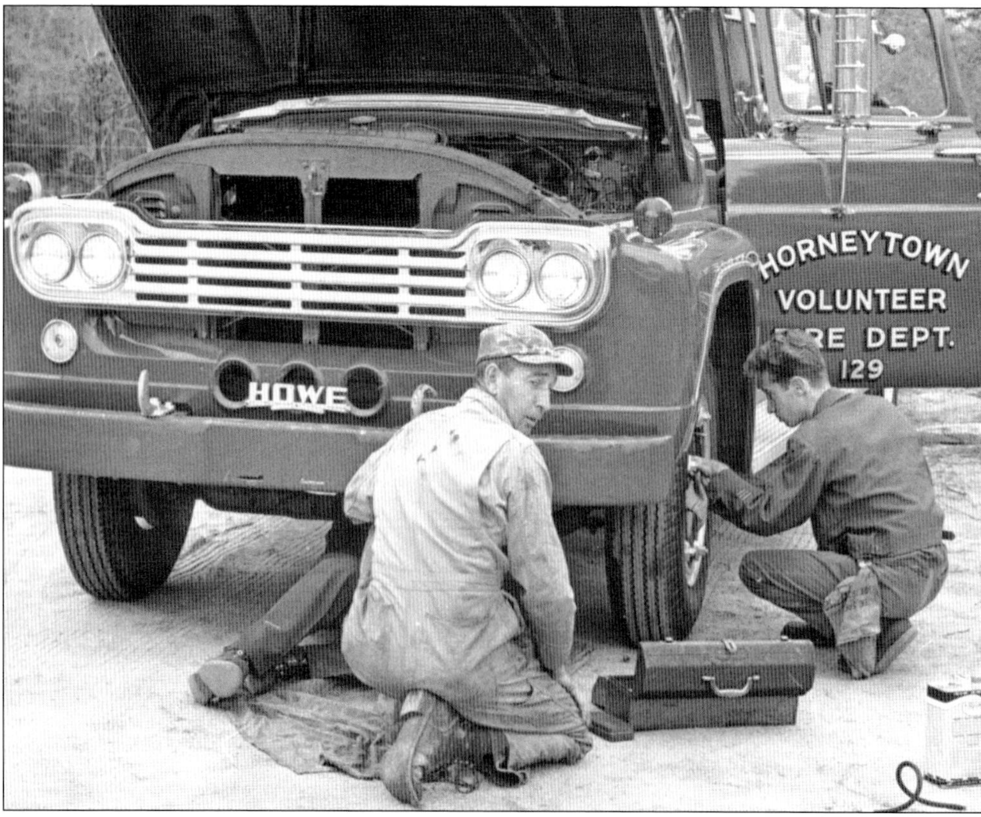

Members of the department are shown here performing maintenance on the Horneytown apparatus in 1960. Repairs and maintenance were often performed by volunteer members at their own expense. (Courtesy of FCFHS.)

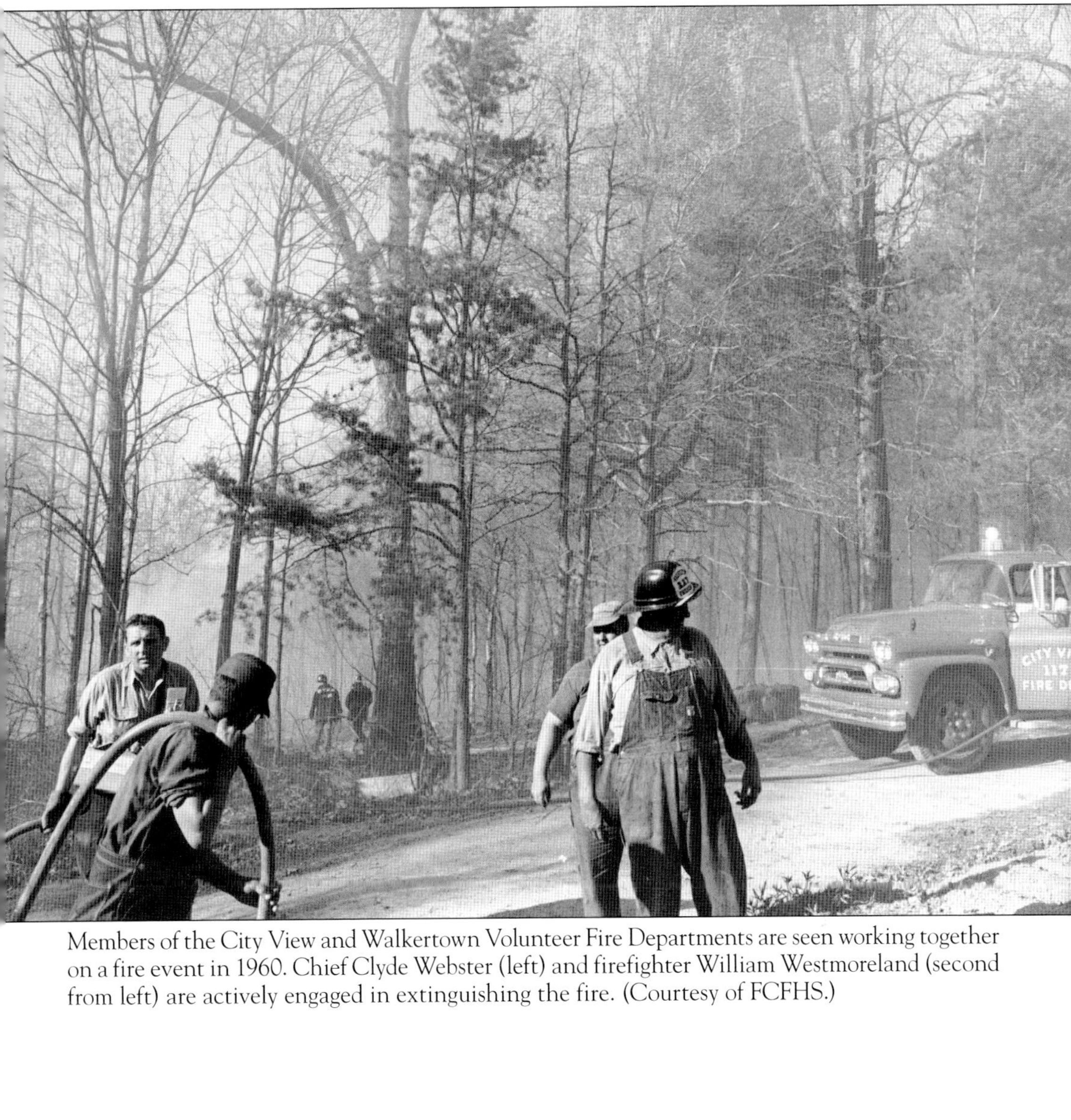

Members of the City View and Walkertown Volunteer Fire Departments are seen working together on a fire event in 1960. Chief Clyde Webster (left) and firefighter William Westmoreland (second from left) are actively engaged in extinguishing the fire. (Courtesy of FCFHS.)

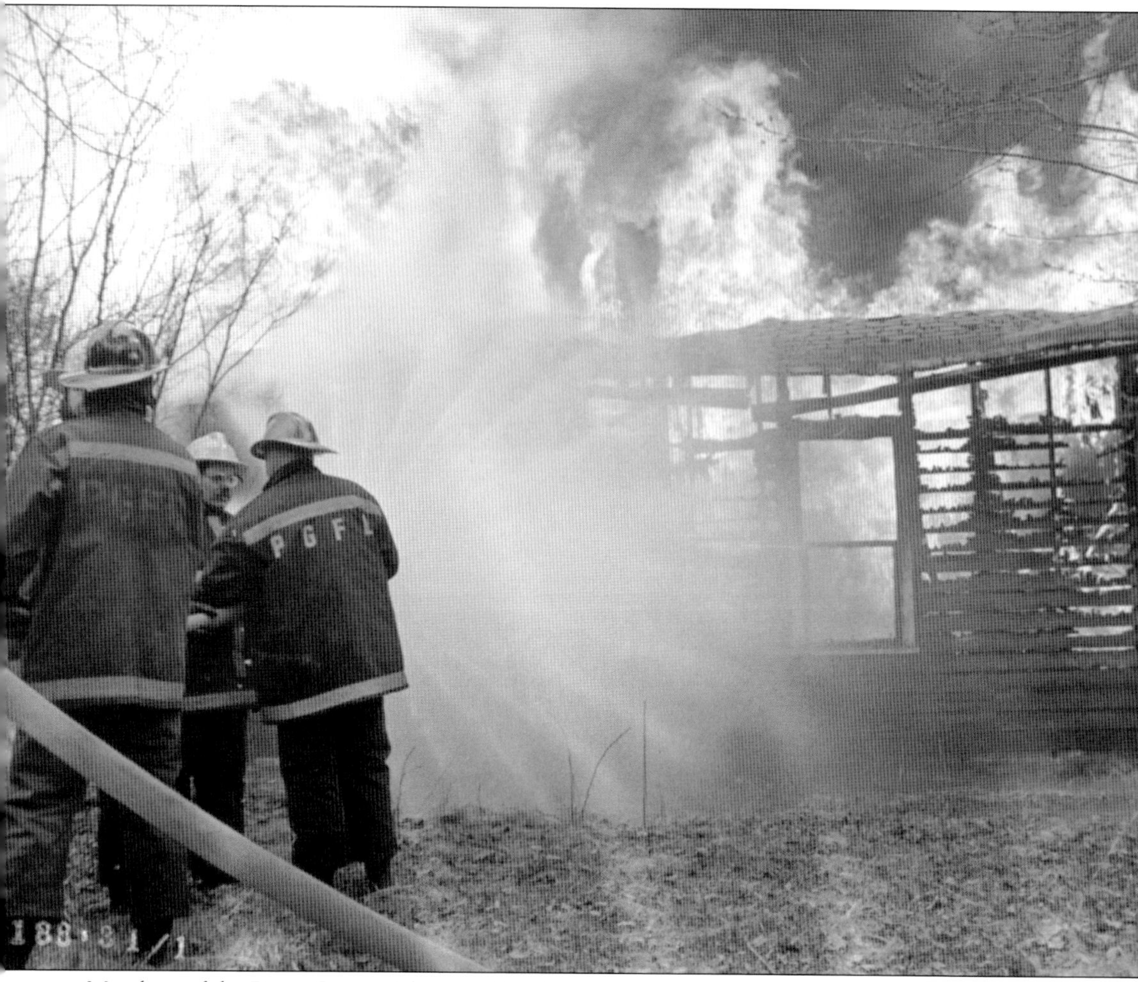

Members of the Piney Grove Volunteer Fire Department are battling a house fire off Vance Road. Although the home was a total loss, no one was injured. House fires were becoming more of a problem in Forsyth County, especially in the cold seasons when fireplaces were used as a source of heat. (Courtesy of FCFHS.)

Three

Forsyth County Fire Department

Before the spring of 1951, there were approximately 75,000 people living in Forsyth County without any fire protection. James G. Hanes, a former chairman of the board of county commissioners, spoke with Chief M.G. Brown about establishing a county fire plan. In 1953, a full-time radio operator was hired, allowing the volunteer fire departments to communicate 24 hours a day, seven days a week. This was an important step in the coordination of volunteer departments. They came to the realization, though, that the cost to fund such a program was too great for the county to finance. This prompted Wally Dunham, owner of the Zinzendorf Hotel and a county commissioner who had a particular interest in rural fire protection, to petition to get the needed funding. Commissioner Dunham spurred county support for a Forsyth County Fire Department that would respond to fires and help coordinate the efforts of the local volunteer fire departments.

The city of Winston-Salem at that time had eight fire stations, so the county commissioners suggested the county department be station No. 9. The Forsyth County commissioners later established a fire commission to organize the development of rural fire protection efforts. The volunteer fire department chiefs in the county requested the Forsyth County Fire Department create a standard for firefighting training activities for all departments. Reece Baugess was appointed fire marshal in 1955, making him the first fire marshal in the state of North Carolina.

In 1956, the county fire departments incorporated the Forsyth County Volunteer Firemen's Association to assist in coordinating fire prevention programs, developing training curriculums, and organizing fundraisers. This also led to the communications division (911) in Forsyth County. In 1963, a station was built at Smith Reynolds Airport to house the communications headquarters. Forsyth County volunteer fire departments were the only stations in the state to have their own radio network at the time. The communications division was formed in 1964. It was the first of its kind in the southeastern United States. Much of the cost to establish a radio system in Forsyth County came from the Department of Civil Defense fund, which already had a system for emergency operations.

The Forsyth County Fire Department has garnered praise from across the country since its humble beginnings. Many rural fire departments have used the Forsyth model to create their own programs.

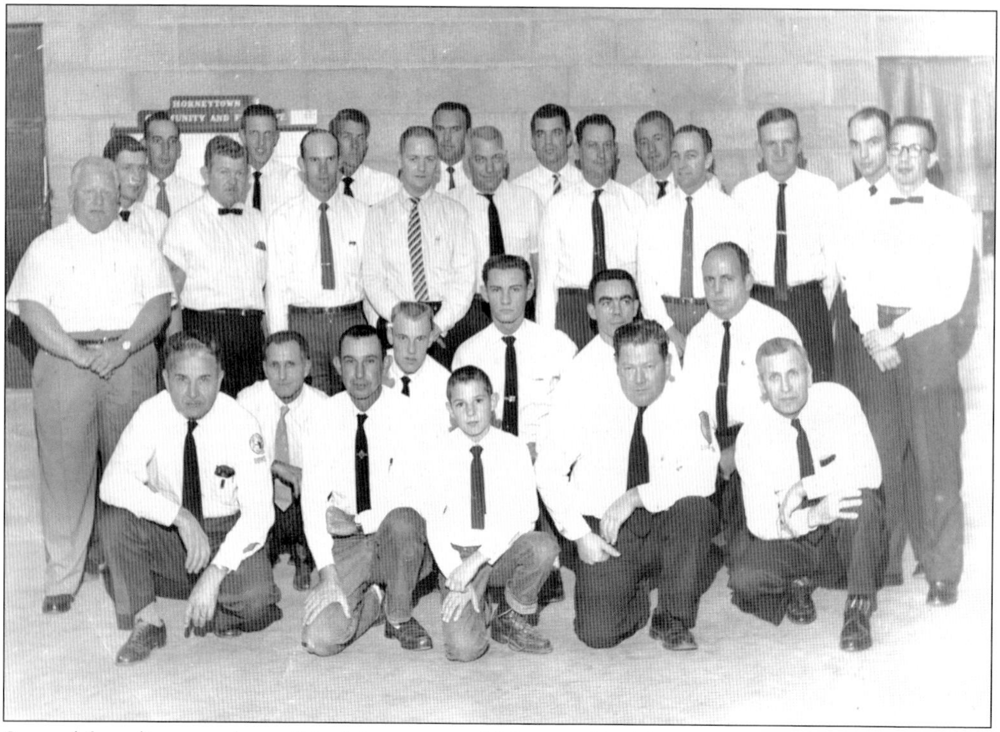

Several founding members of the Horneytown Volunteer Fire Department pose for a photograph in the department bay on Horneytown Road in 1960. (Courtesy of FCFHS.)

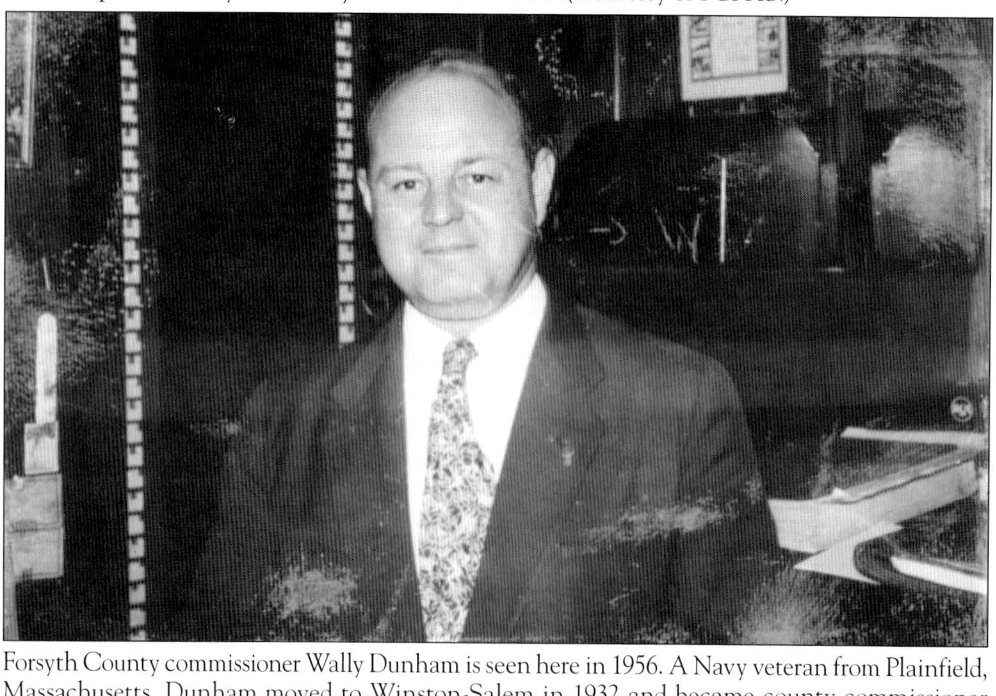

Forsyth County commissioner Wally Dunham is seen here in 1956. A Navy veteran from Plainfield, Massachusetts, Dunham moved to Winston-Salem in 1932 and became county commissioner in 1950. Commissioner Dunham served the county from 1950 to 1960 and was instrumental in establishing many rural volunteer fire departments. (Courtesy of FCFHS.)

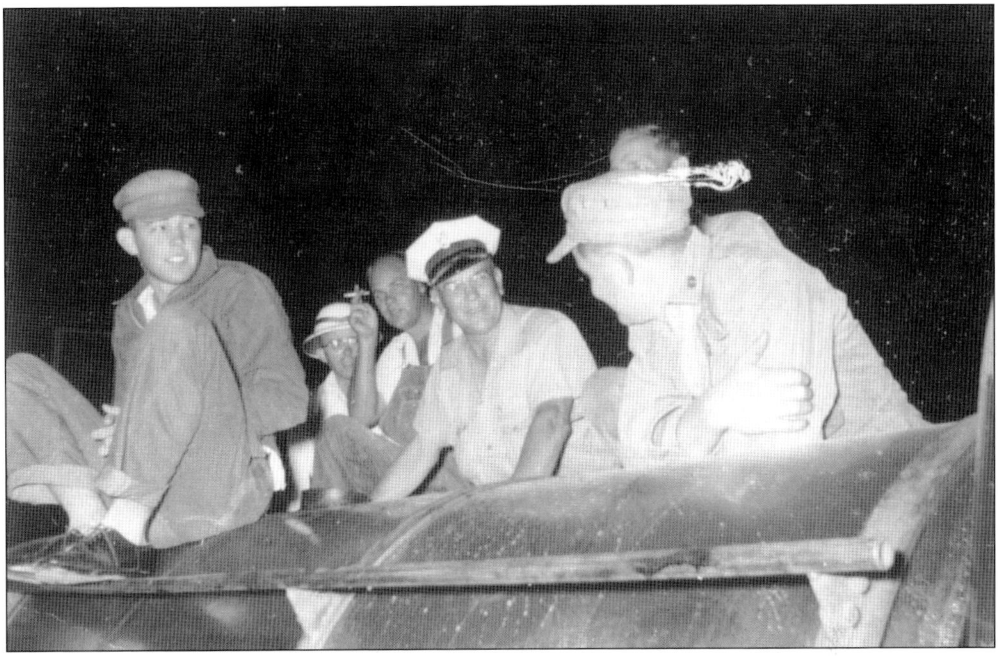

Fire department volunteer members are shown here training on water tank operations in 1960. The Forsyth County Fire Department assisted in training efforts for all volunteer departments. (Courtesy of FCFHS.)

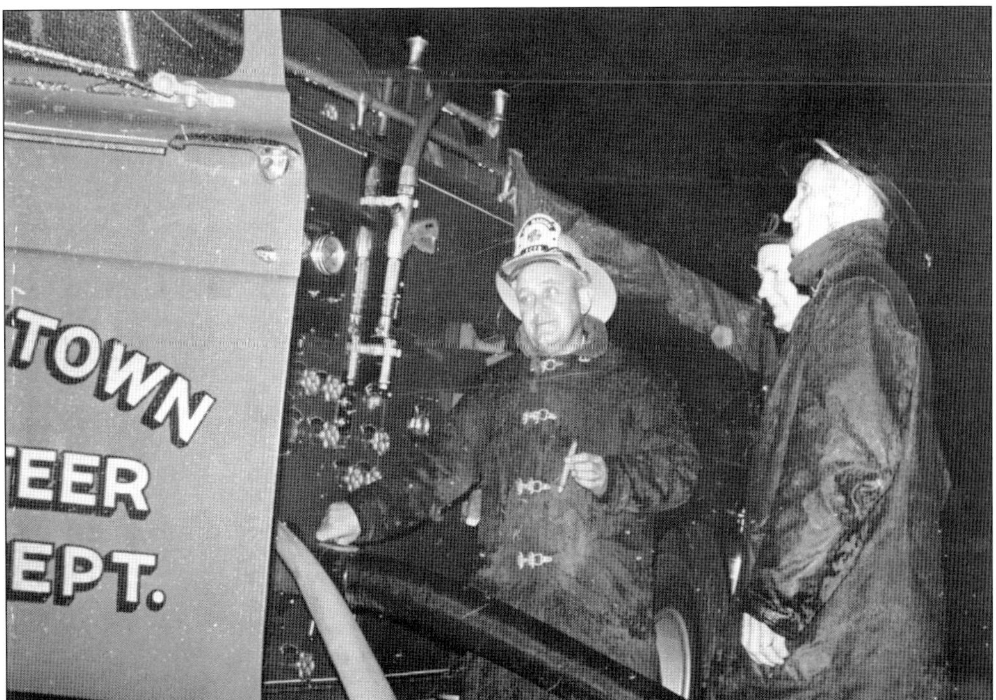

Forsyth County fire marshal Reese Bauguess is seen assisting members of the Horneytown Volunteer Fire Department. (Courtesy of FCFHS.)

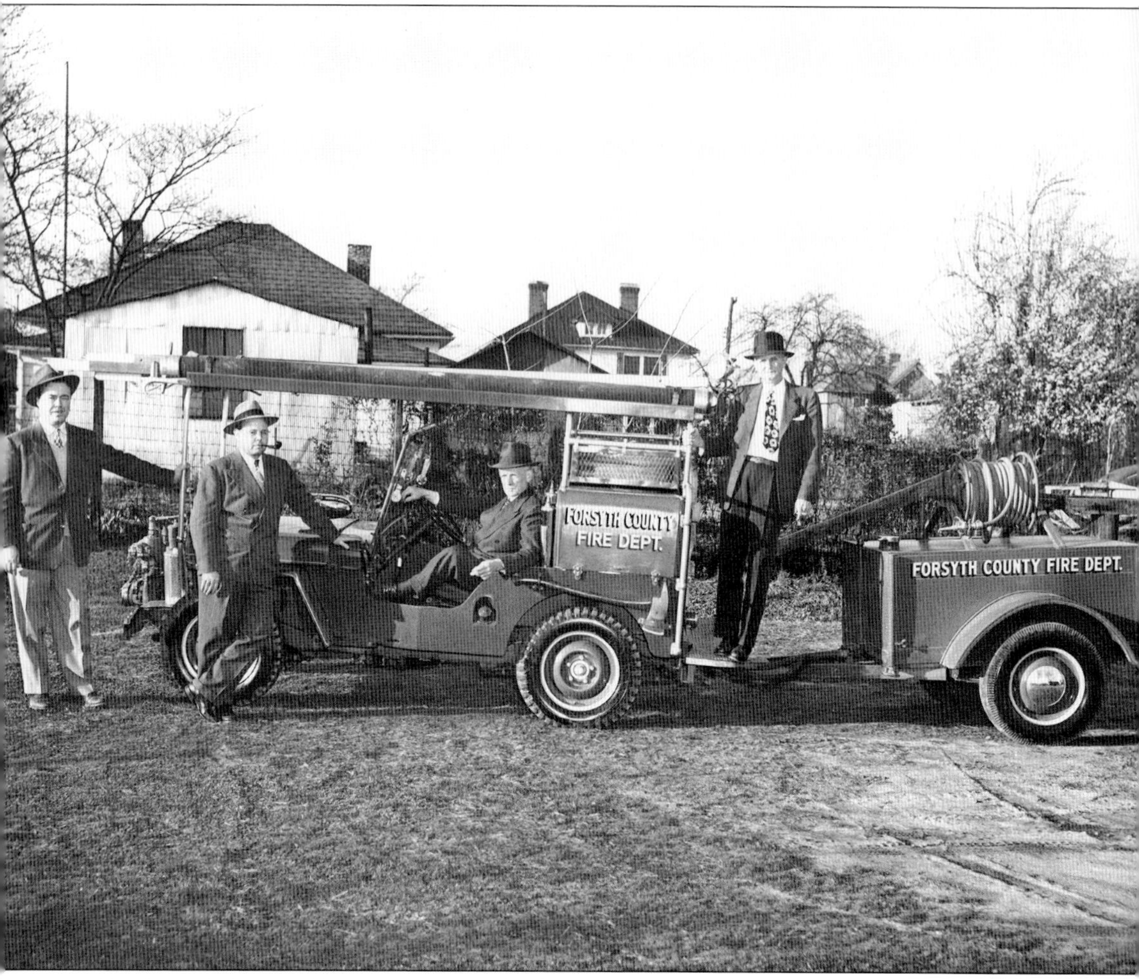

The Forsyth County Fire Department 109 (Station 9) jeep is shown with county commissioners (from left to right) William Simpson, Wally Dunham, Dr. D.C. Speas, and Carl Ogburn in 1951. The 1944 jeep featured a front-mounted pump and trailer with a water tank and booster hose reel. It would respond all over Forsyth County to assist in firefighting efforts. (Courtesy of FCPL.)

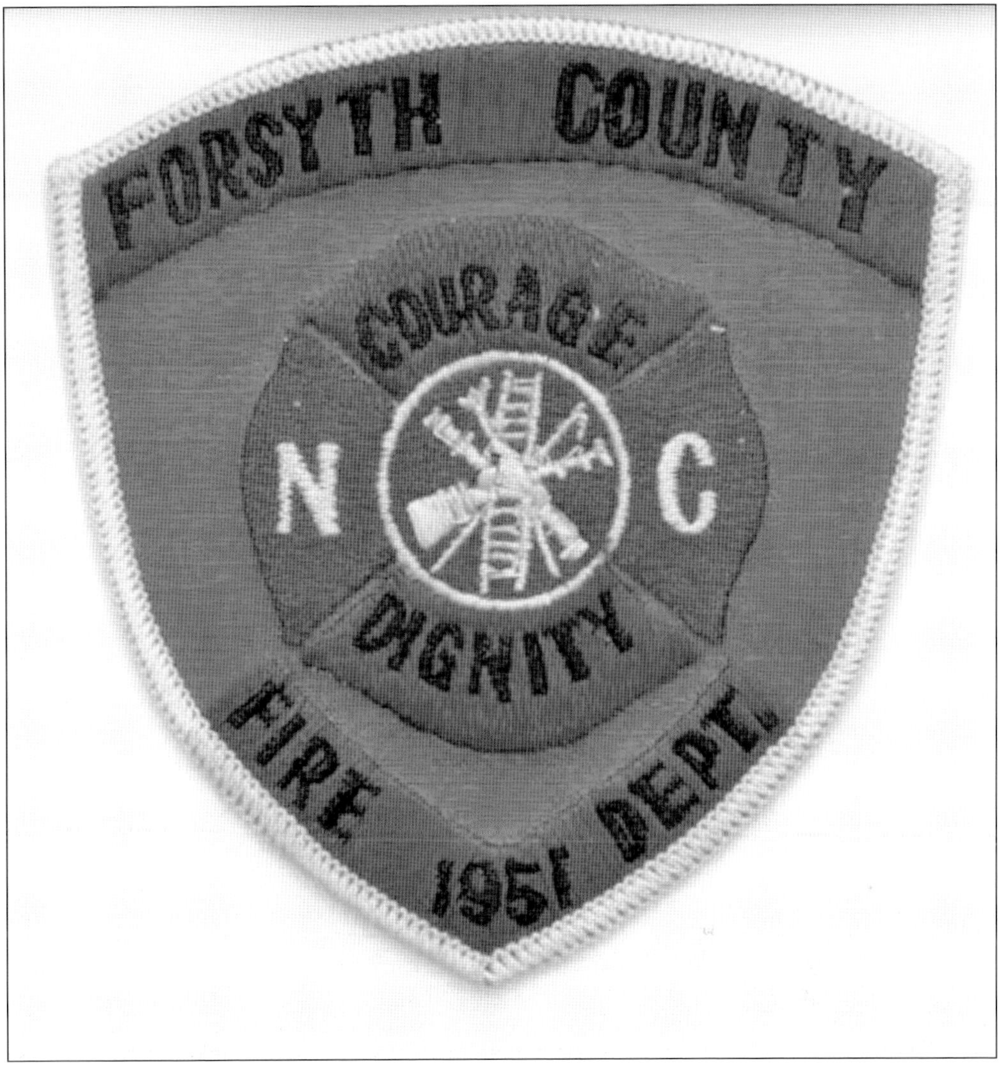

This is the original patch of the Forsyth County Fire Department from 1951. The patch featured a traditional Maltese cross with the words courage and dignity. These words hold true today as courage is a characteristic that all firefighters must have. Dignity is necessary to do their job with pride and love for the community. (Courtesy of FCFHS.)

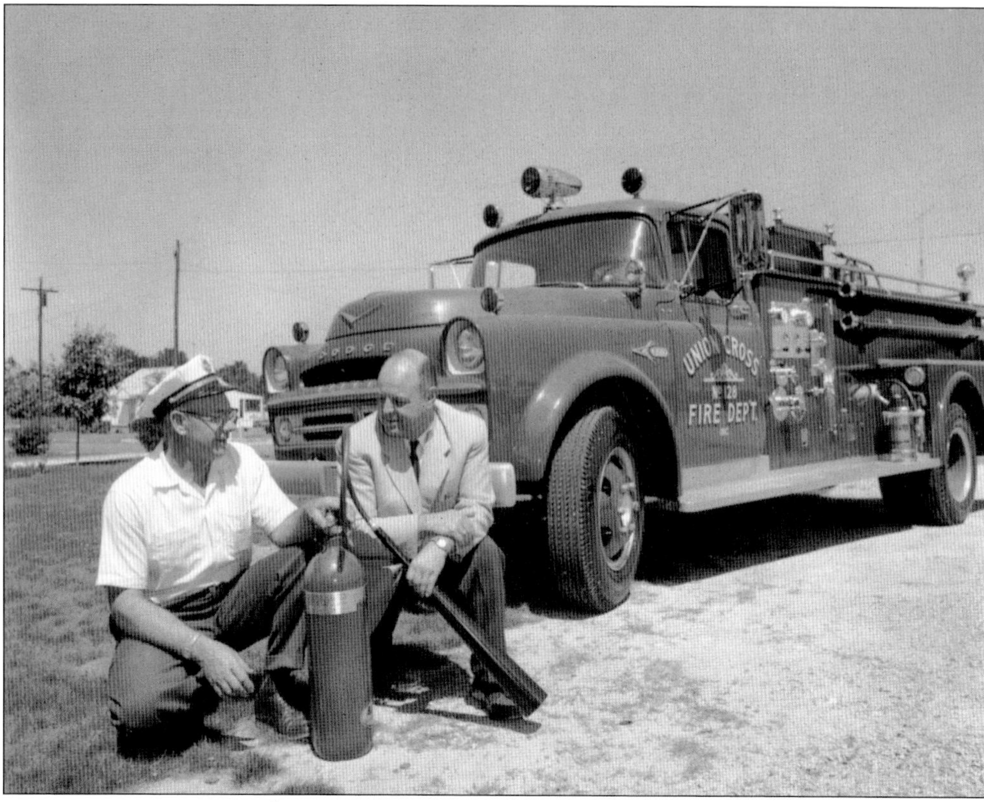

County commissioner Wally Dunham is seen with Union Cross volunteer fire chief Oscar Weavil looking over a CO2 extinguisher in 1958. CO2 extinguishers were a beneficial tool for fire departments as they allowed for a nonconductive way to extinguish switchboard fires. (Courtesy of FCFHS.)

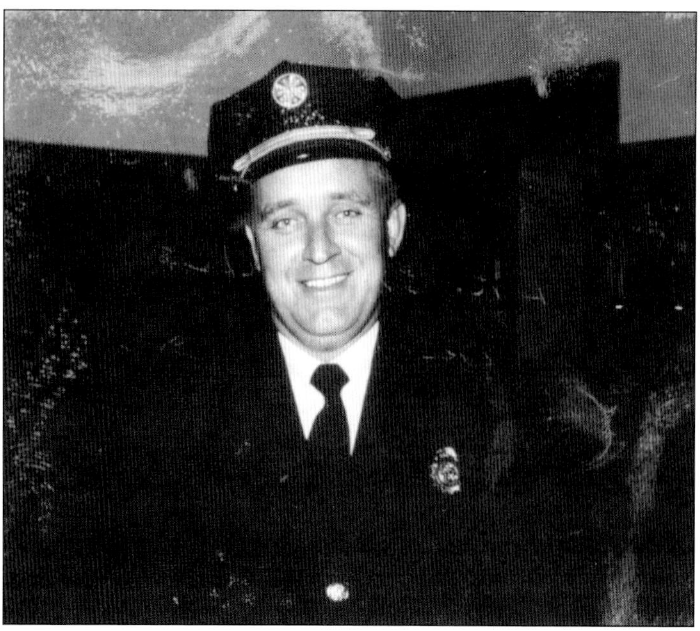

Reese Bauguess was appointed the first fire marshal in the state of North Carolina. This replaced the position of county fire chief in 1954. Bauguess was originally from Virginia and relocated to North Carolina after an early career in tobacco marketing. He later established several volunteer and combination fire departments in the state. (Courtesy of FCFHS.)

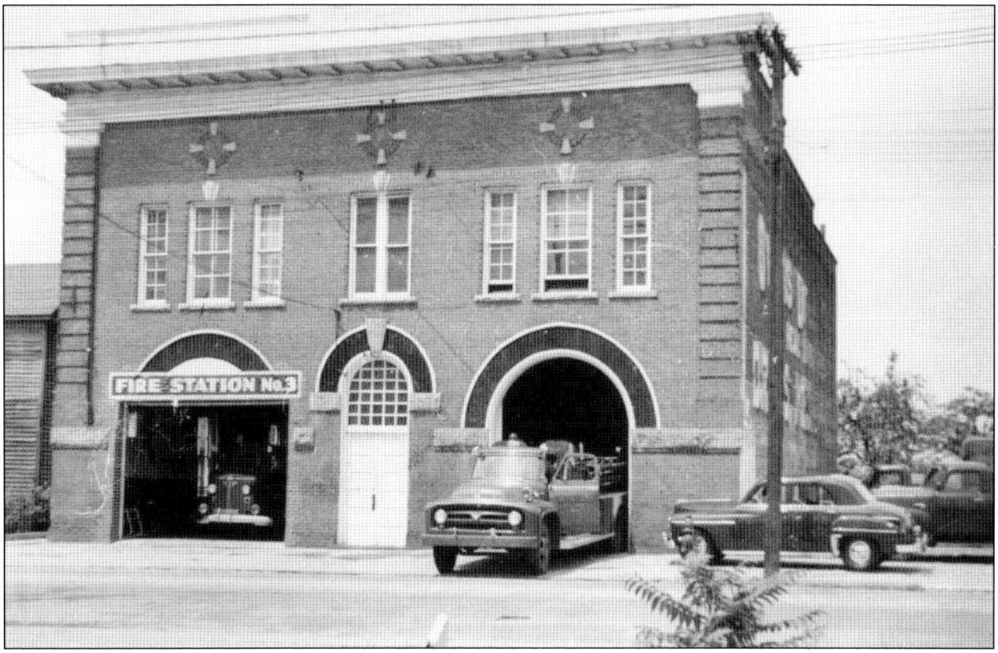

The Forsyth County Bean Chevy 109 truck was originally housed at the Winston-Salem Fire Department Station No. 3 on North Liberty Street in 1955. (Courtesy of FCFHS.)

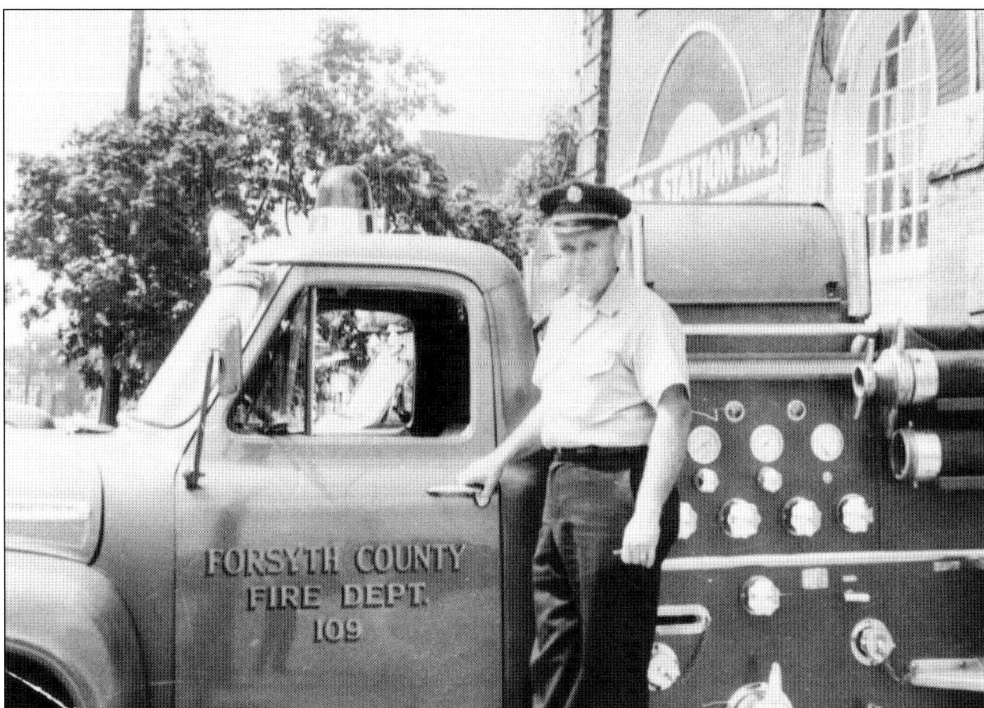

Firefighter F.W. Broadstreet is standing next to the 1952 Bean Chevy 109 truck at the Winston-Salem Fire Department Station No. 3 in 1954. (Courtesy of FCFHS.)

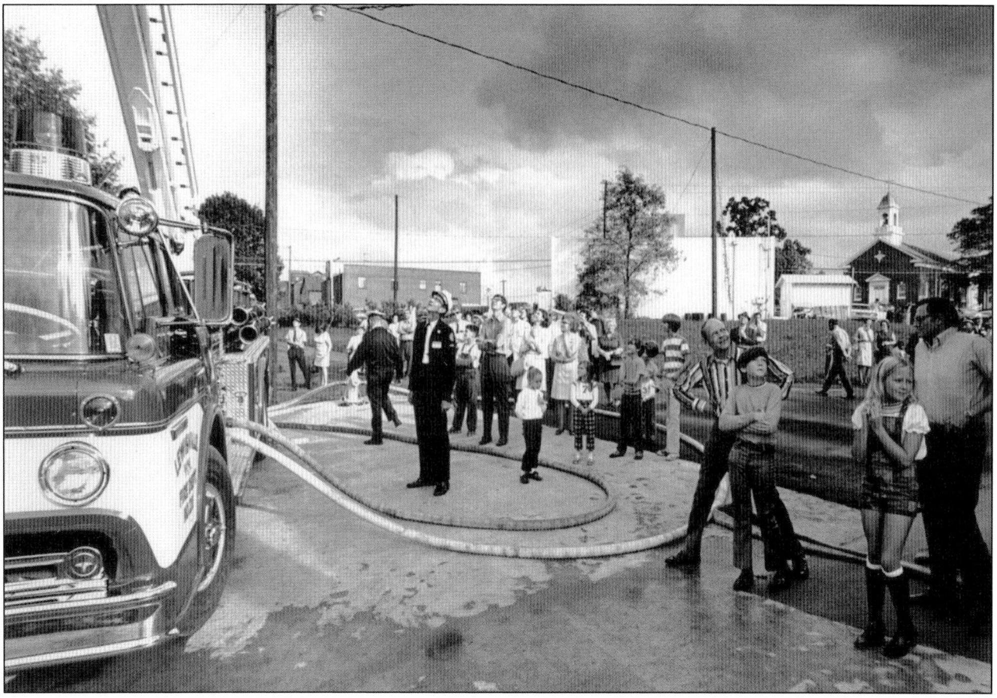

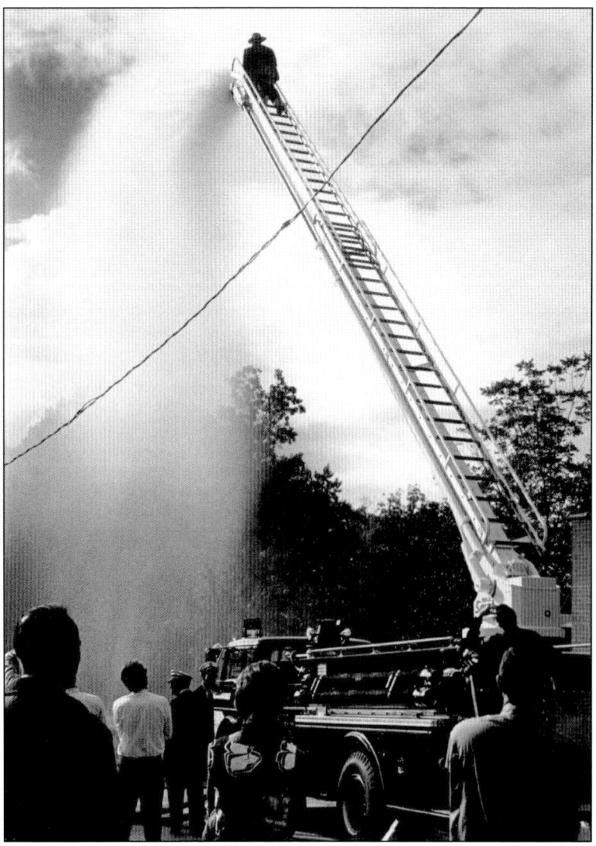

The Lewisville Volunteer Fire Department is shown on the scene of a large fire event. The 1971 Pierce 50-foot Telesquirt aerial truck is also shown. The Lewisville Volunteer Fire Department was the first volunteer department in the state to purchase an aerial unit, which allowed for greater access for hose lines to extinguish structural fires. (Both, courtesy of FCFHS.)

This was the first Forsyth County Volunteer Fire Department patch. Patches were used to identify which department firefighters were affiliated with. (Courtesy of FCFHS.)

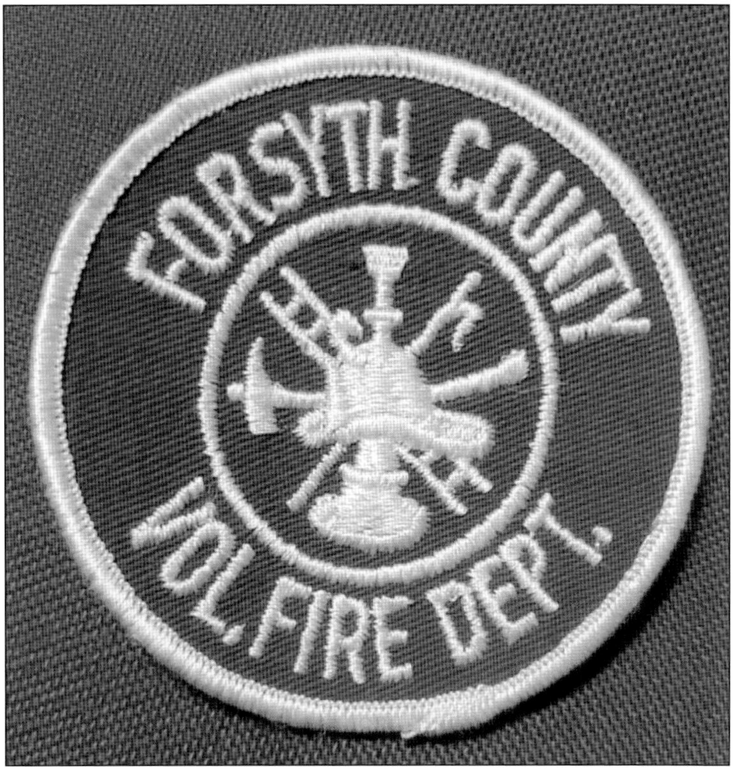

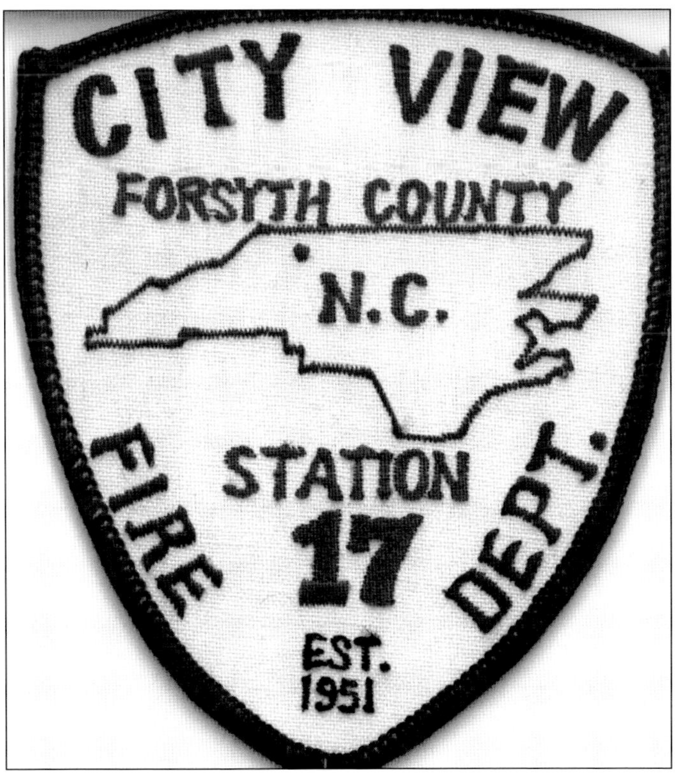

Each department wore its Forsyth County volunteer patch with the identifying department name underneath it. Individual departments later created their own patches. The City View Fire Department patch is pictured here. (Courtesy of FCFHS.)

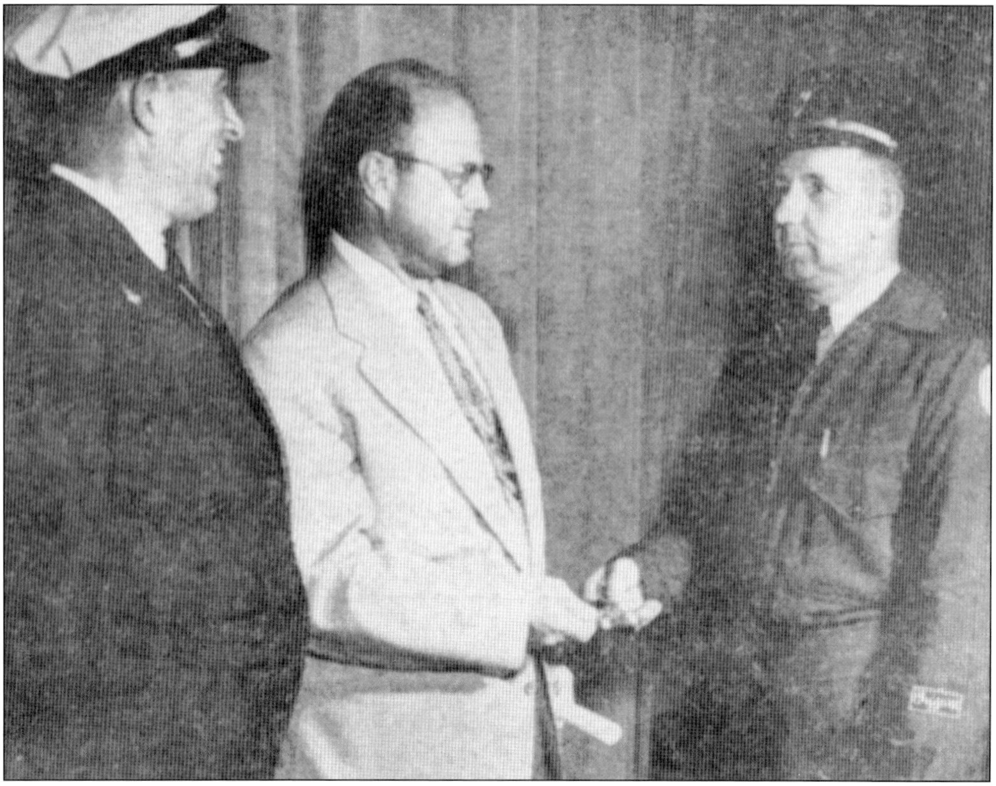

County commissioner Wally Dunham (center) awards a badge to Griffith Fire Department member Ralph Zimmerman (right) for completing the firefighting course established by Forsyth County. Charles Fox (left) was the instructor of the class. (Courtesy of FCFHS.)

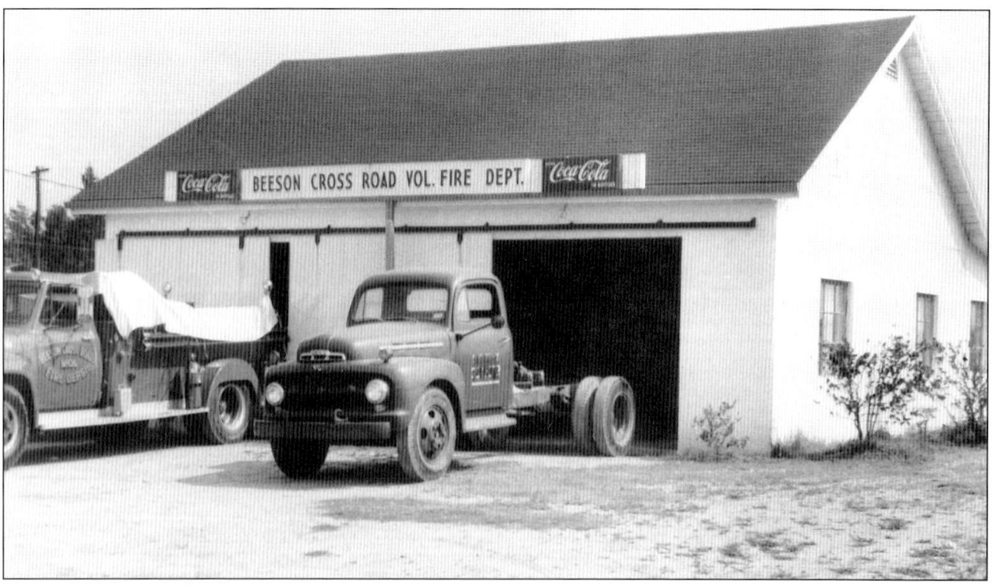

Pictured here is the Beeson Cross Roads Volunteer Fire Department building in 1955. The department was located on land owned by Robert Joyce with a lease of 99 years. Edgar Idol was elected chief of the department in 1954. (Courtesy of FCFHS.)

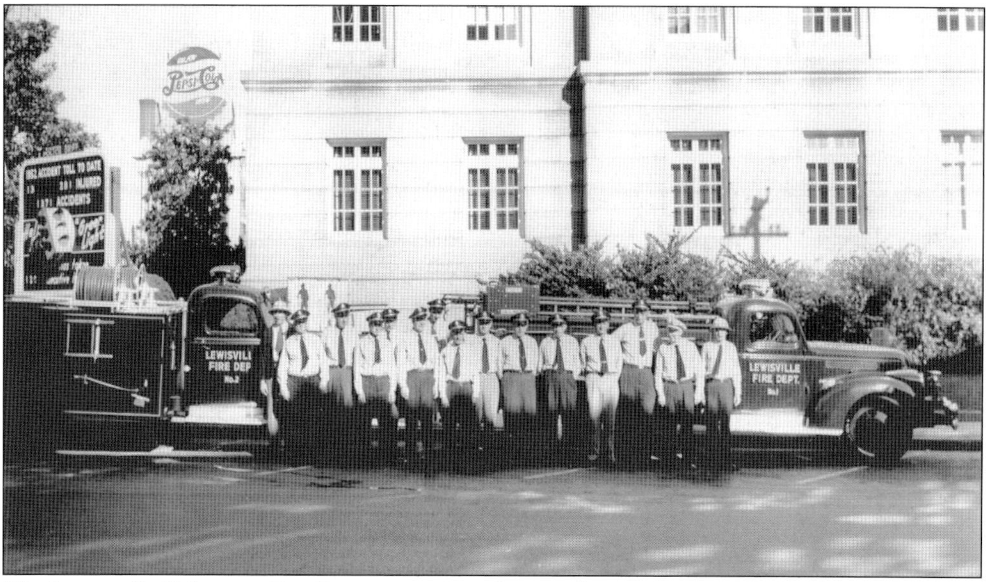

The Lewisville Volunteer Fire Department members and apparatus are seen here in front of the Winston-Salem Courthouse in 1959. (Courtesy of FCFHS.)

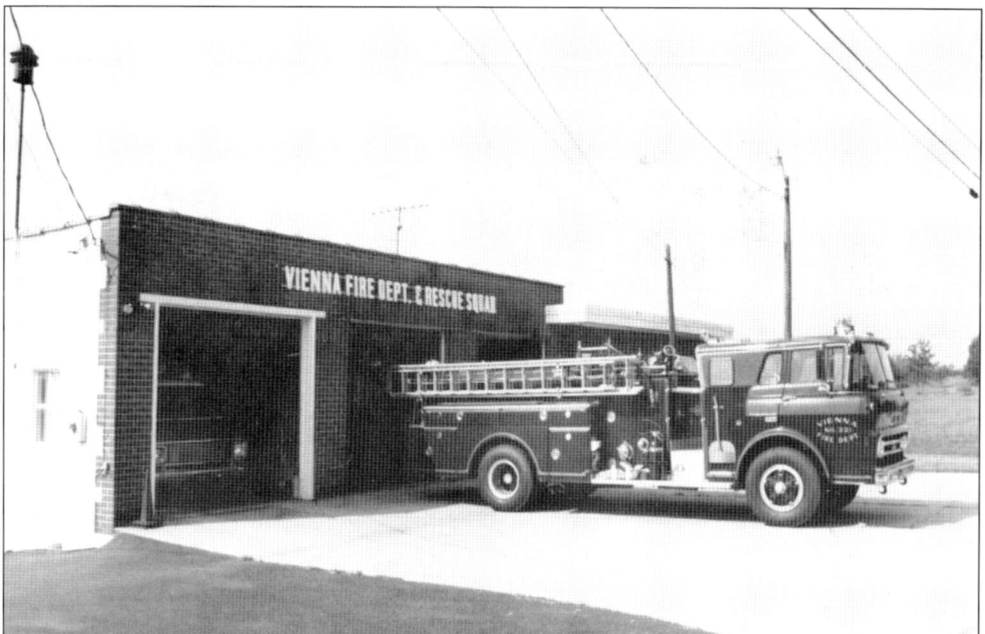

The Vienna Volunteer Fire Department was established in 1954. This department was proposed by the Vienna Civic Club at the suggestion of Marvin Beroth. Prior to this department, Lewisville and Mount Tabor volunteer departments were responsible for the area's fire protection. (Courtesy of FCFHS.)

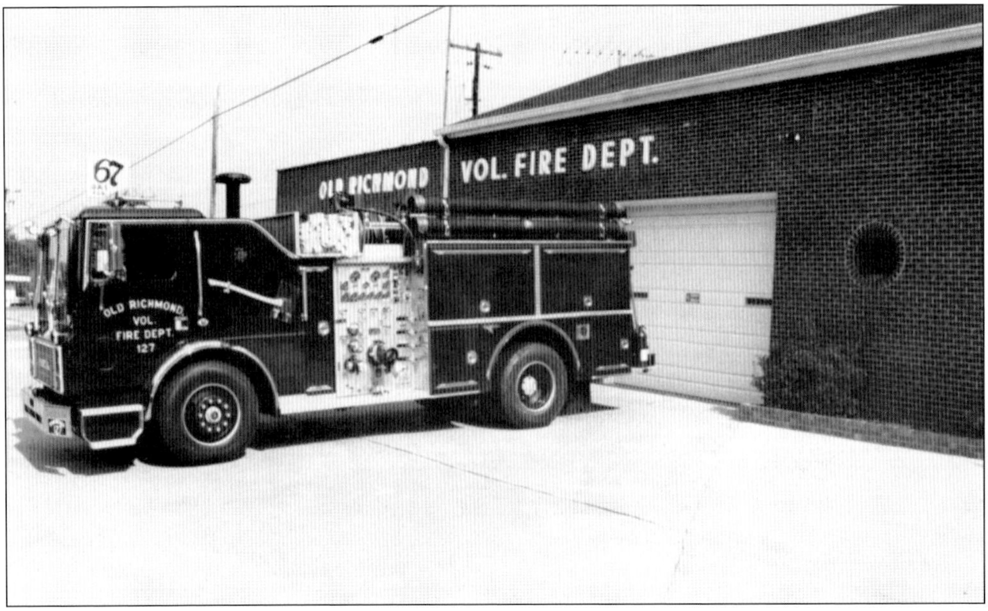

The Old Richmond Volunteer Fire Department was established in 1956. The Old Richmond Grange took the first steps in forming the department in 1954. After the charter was granted, Marvin Doub was appointed the fire chief for the department. The majority of the funds to purchase apparatus and equipment were raised by the Ladies Auxiliary. (Courtesy of FCFHS.)

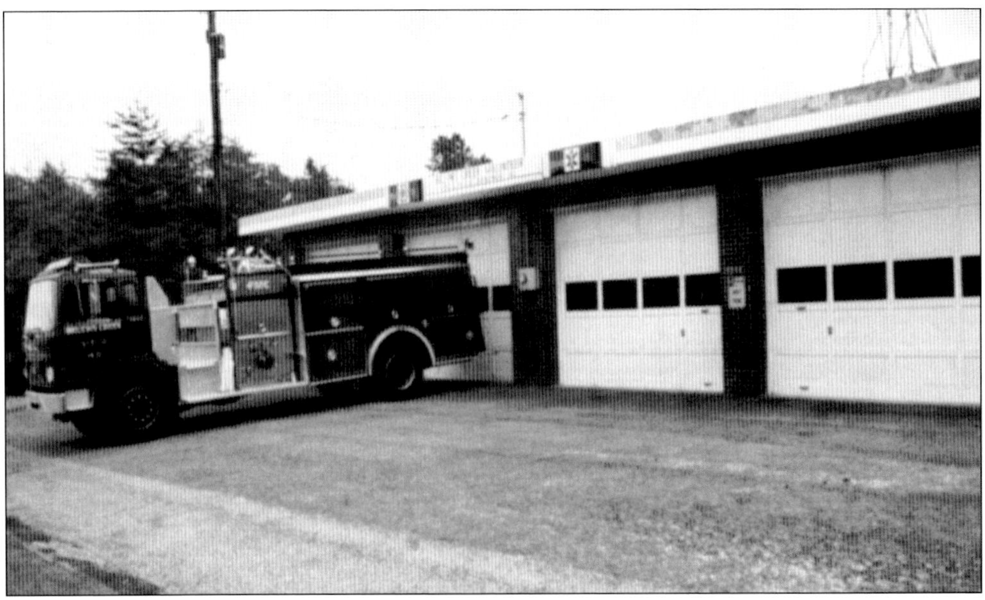

The Belews Creek Volunteer Fire Department was established in 1962. The first meeting of this department was held at the Goodwill Community Center. In 1972, a steering committee recommended that a station be built at Crims Cross Roads. (Courtesy of FCFHS.)

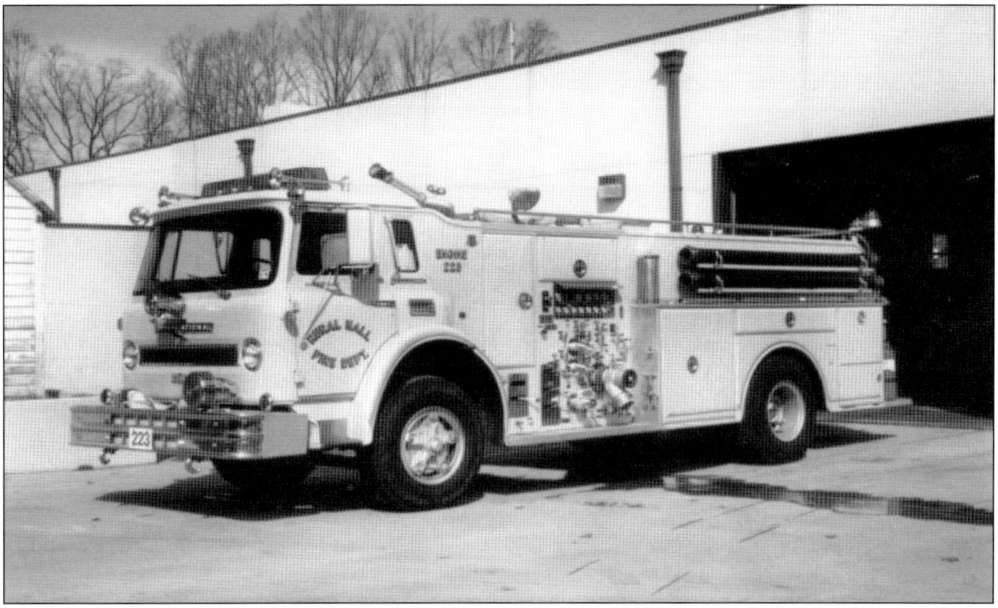

Shown is the Rural Hall Fire Department engine No. 223, a 1974 International/Howe with a 1,000-gallons-per-minute pump. The Rural Hall Volunteer Fire Department was established in 1939 and is credited as the first volunteer fire department in Forsyth County. (Courtesy of FCFHS.)

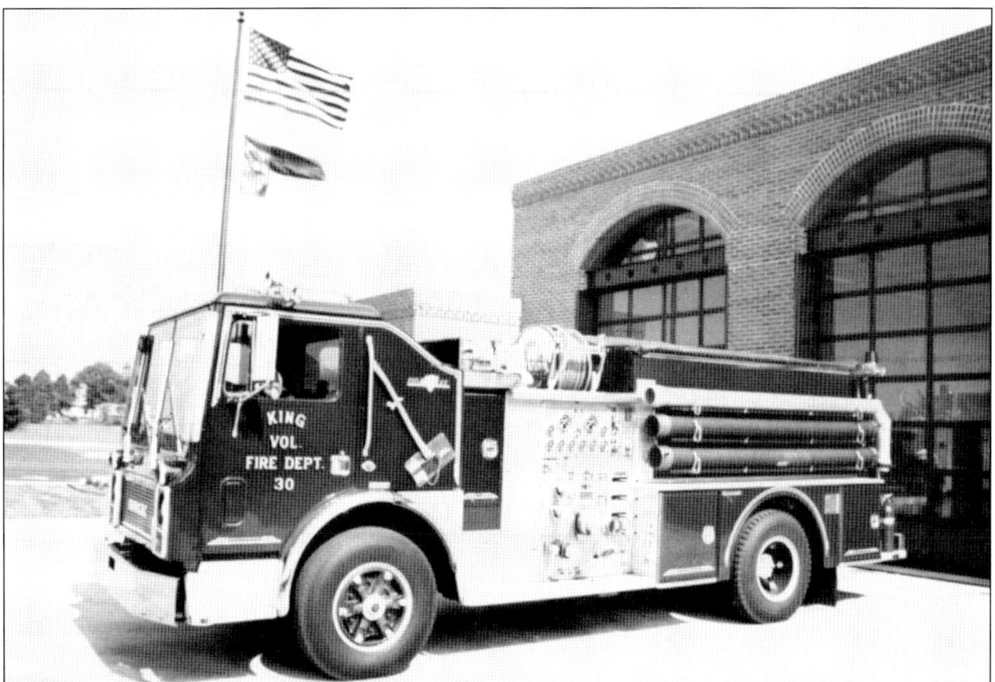

The King Volunteer Fire Department was formed in 1955. The Lions Club decided to form a committee to establish a department for the town. Prior to the committee formation, the town's only fire protection was a 50-pound wheeled extinguisher that was rolled out for use in the daytime. (Courtesy of FCFHS.)

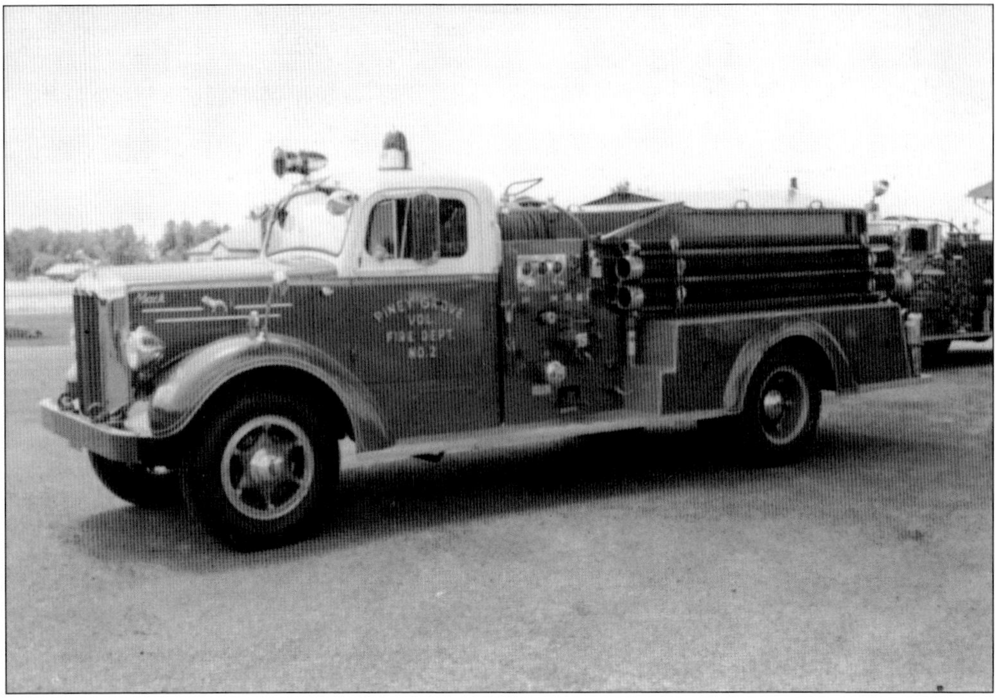

A meeting was held at the Piney Grove United Methodist Church in 1967 to create the Piney Grove Volunteer Fire Department. The department's 1953 Mack fire apparatus is pictured here. (Courtesy of FCFHS.)

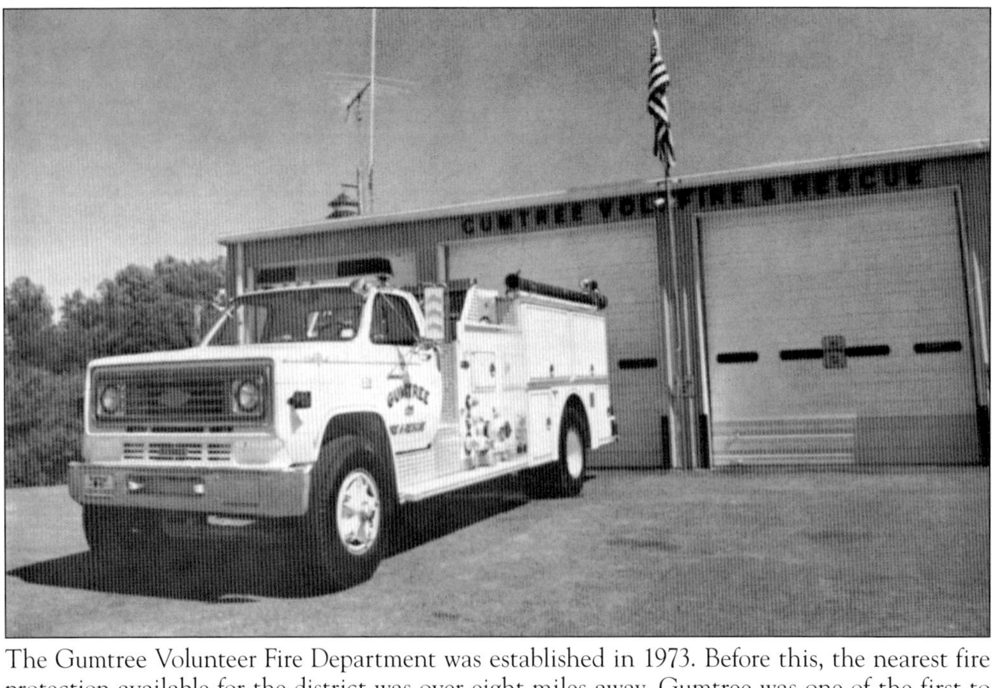

The Gumtree Volunteer Fire Department was established in 1973. Before this, the nearest fire protection available for the district was over eight miles away. Gumtree was one of the first to provide an ambulance assist program for its district. In 1977, the department became Gumtree Volunteer Fire and Rescue. (Courtesy of FCFHS.)

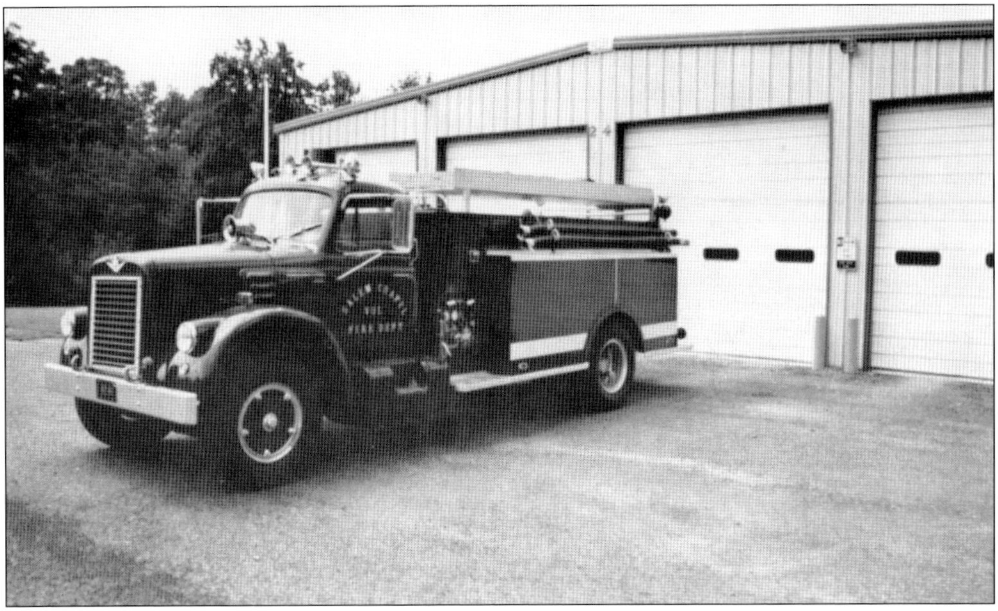

During the annual meeting of the Walkertown Volunteer Fire Department in January 1982, a decision was made to create some form of fire protection for the Salem Chapel area. Another meeting, to establish that volunteer department, was held at Fulp Moravian Church, and the Salem Chapel Volunteer Fire Department was created later in 1982. The Walkertown Volunteer Fire Department donated its first fire truck. (Courtesy of FCFHS.)

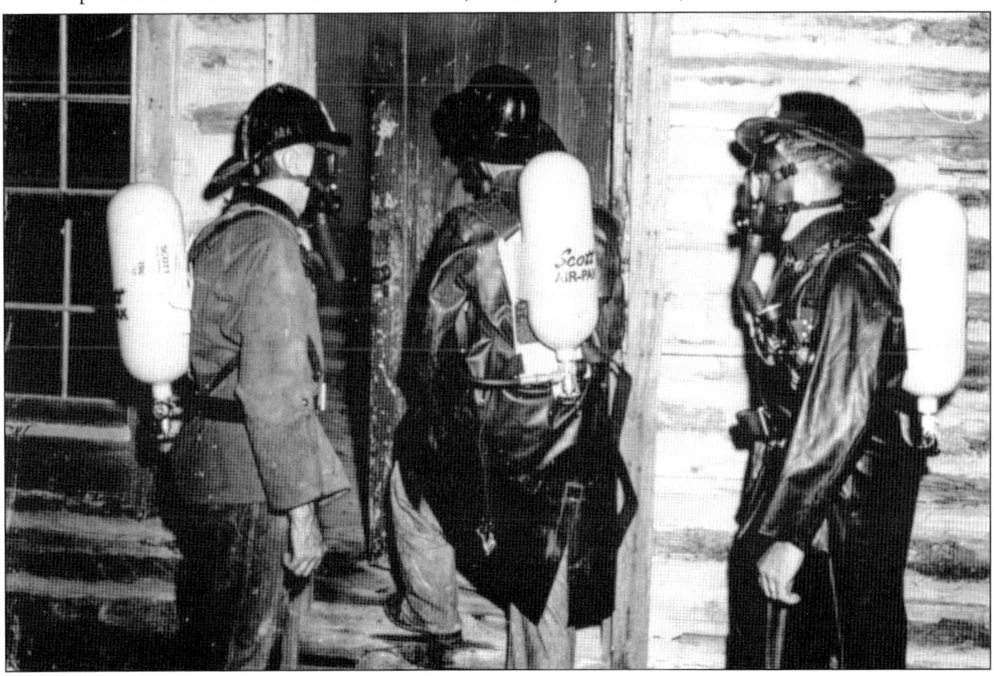

Lewisville Volunteer Fire Department members are using new Scott Air-Pak bottles on a fire scene in 1958. Air-paks were an essential tool of the trade for firefighters. Fundraising campaigns allowed for the purchase of more air-paks for the firefighters, allowing them to go into structures and fight fires without injuring their airways. (Courtesy of FCFHS.)

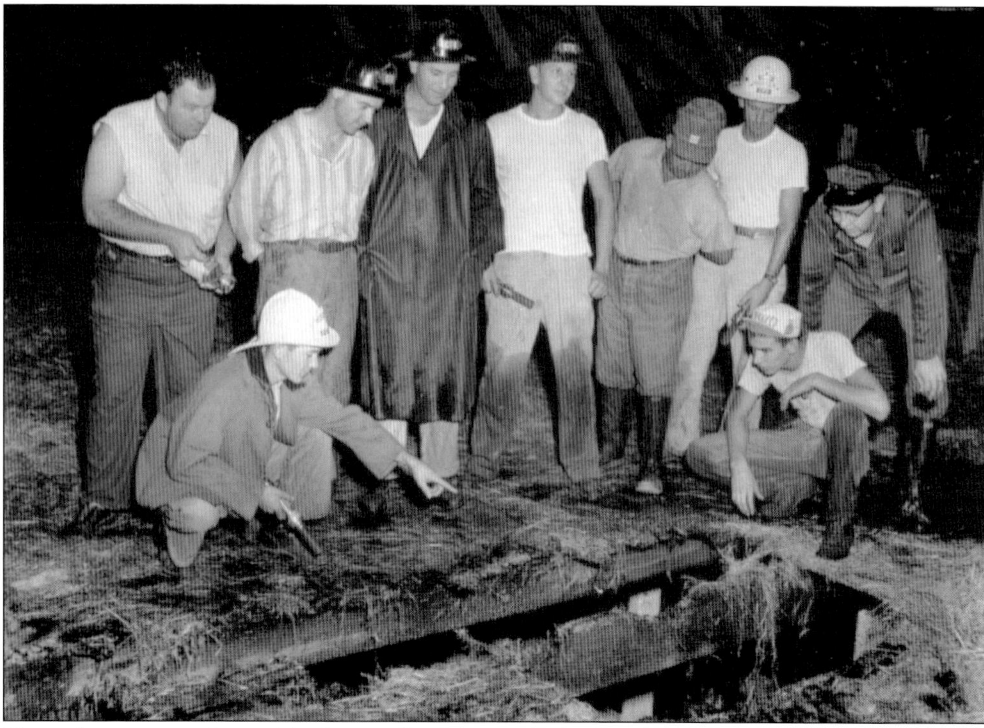

Shown here are members of the Lewisville Volunteer Fire Department learning about fire behavior and control after a fire event. Training and learning more about the profession is necessary for all firefighters. (Courtesy of FCFHS.)

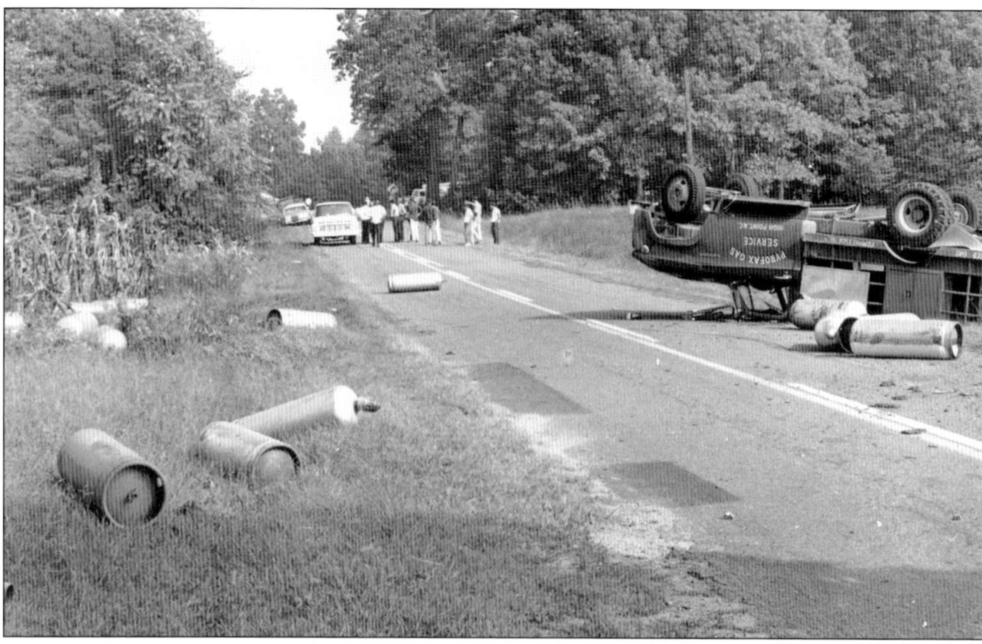

Here, a vehicle carrying cylinders has overturned on a rural road in the Horneytown fire district in 1960. Volunteer members can be seen assessing the scene in the distance. Most volunteer firefighters responded to calls in their own vehicles. (Courtesy of FCFHS.)

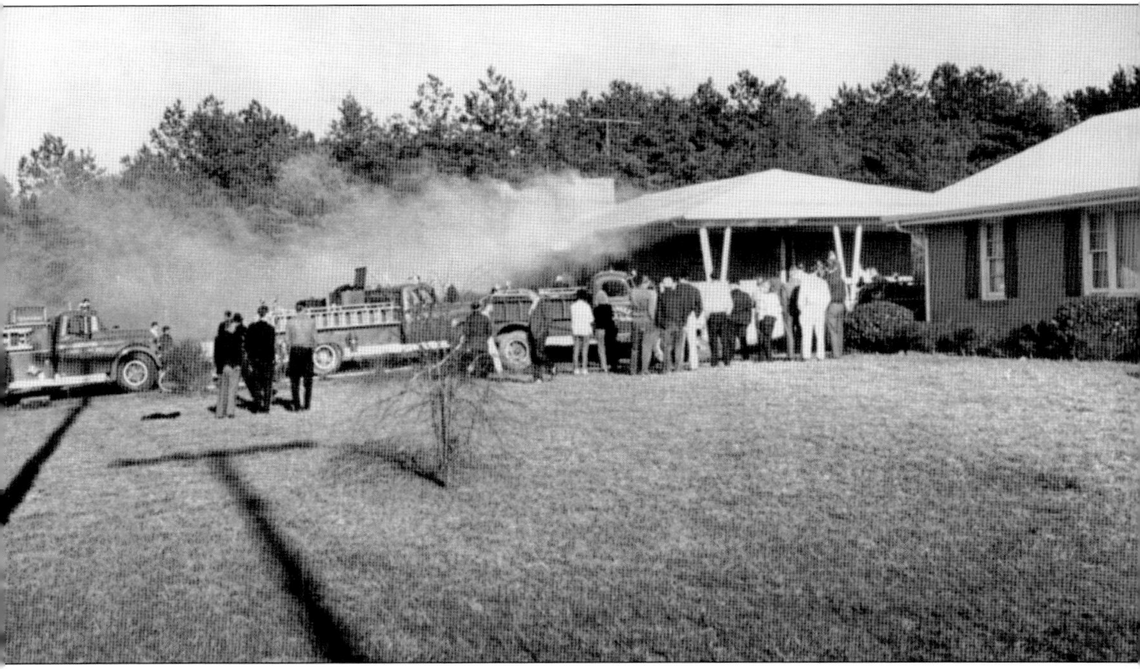

The Forsyth County's 109 truck is being used on a house fire in the City View fire district. The Triangle Volunteer Fire Department apparatus can also be seen. (Courtesy of FCFHS.)

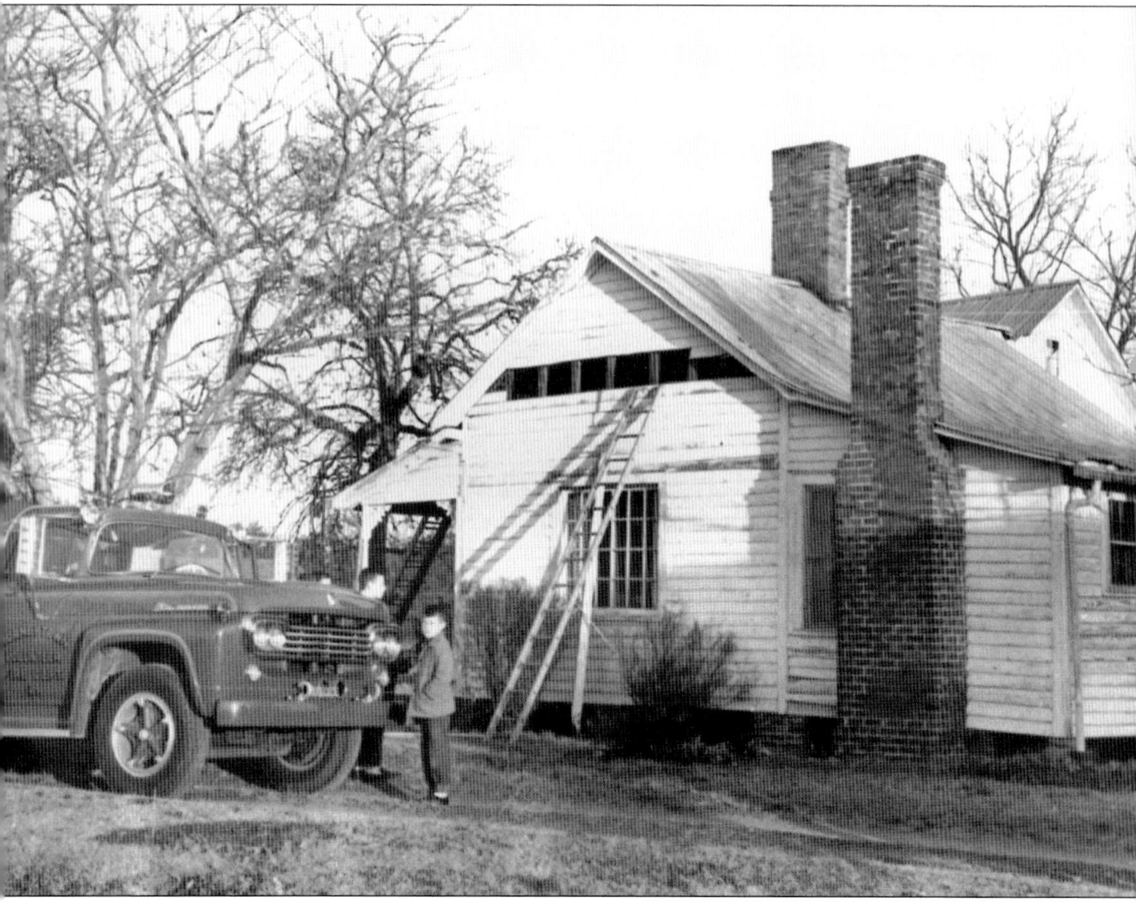

The Horneytown Volunteer Fire Department apparatus is seen here after a fire broke out in the attic of a home. The two young men pictured are most likely children of one of the volunteer members called to extinguish the blaze. (Courtesy of FCFHS.)

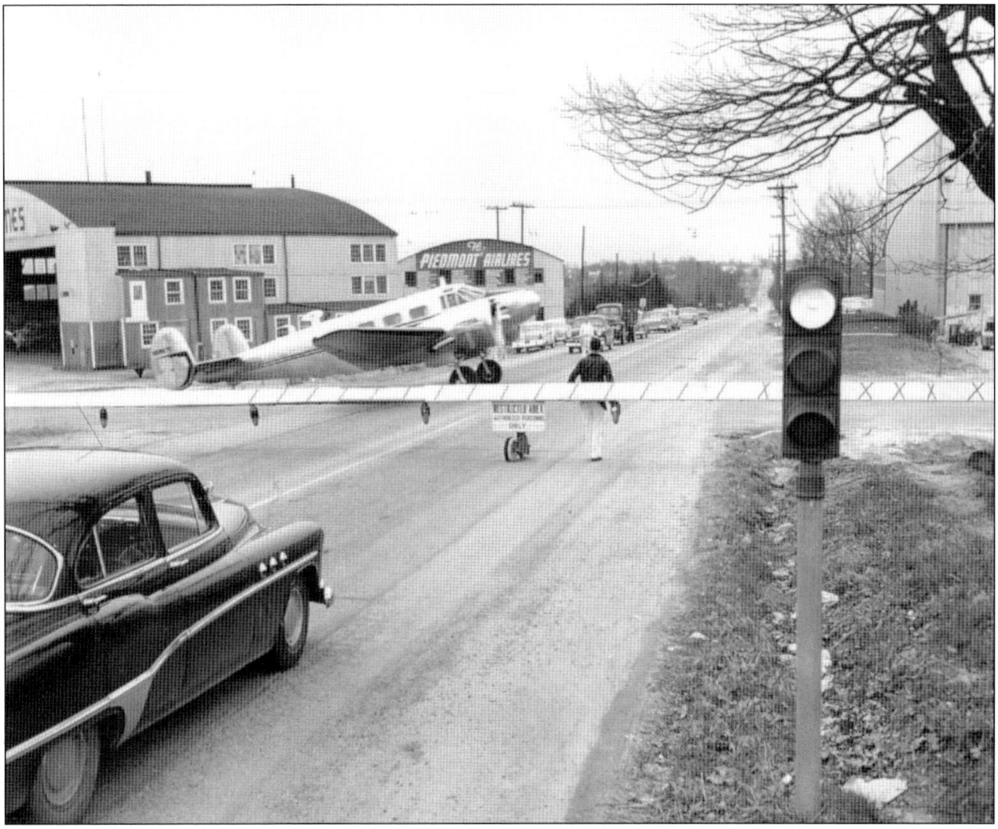

This airplane crossing on Liberty Street in Winston-Salem led to the Smith Reynolds Airport in 1955. (Courtesy of FCFHS.)

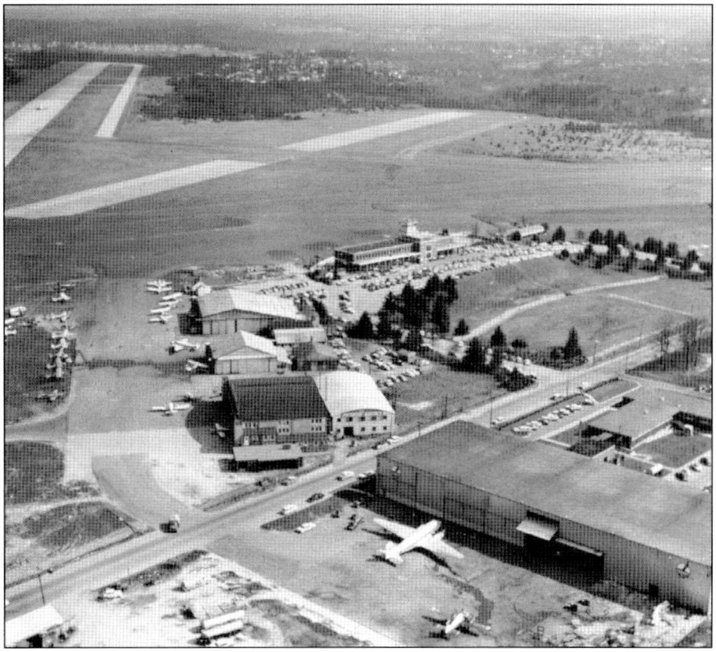

The Smith Reynolds Airport in Winston-Salem was home to many businesses, including Piedmont Airlines. The airport also hosted the Winston-Salem Air Show. It was named after aviator Zachary Smith Reynolds, the youngest son of tobacco mogul Richard Joshua Reynolds. (Courtesy of FCFHS.)

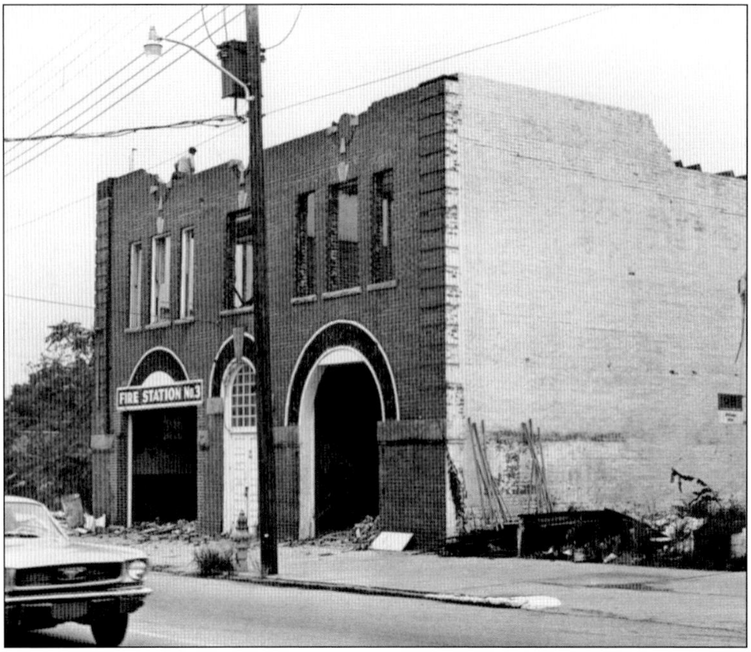

The Winston-Salem Fire Department Station No. 3 was on North Liberty Street and was demolished in 1971. This was where the county's 109 apparatus was housed. (Courtesy of FCFHS.)

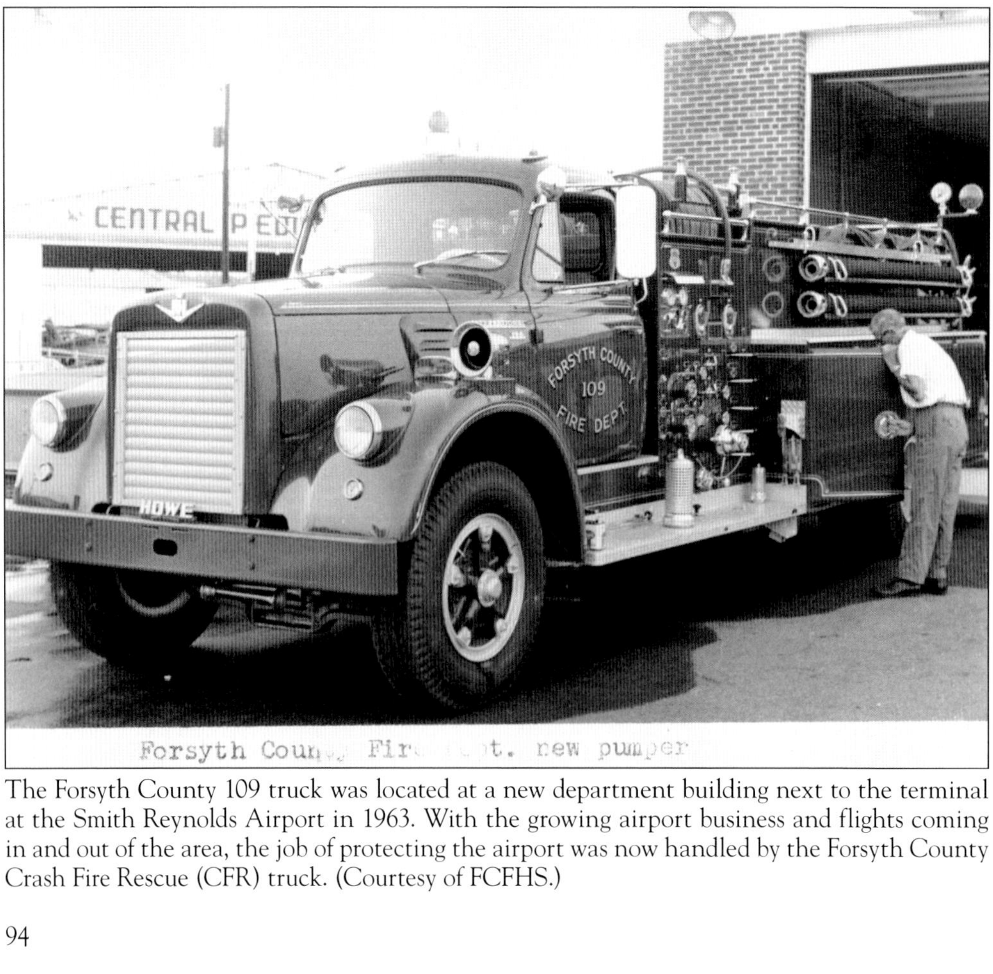

The Forsyth County 109 truck was located at a new department building next to the terminal at the Smith Reynolds Airport in 1963. With the growing airport business and flights coming in and out of the area, the job of protecting the airport was now handled by the Forsyth County Crash Fire Rescue (CFR) truck. (Courtesy of FCFHS.)

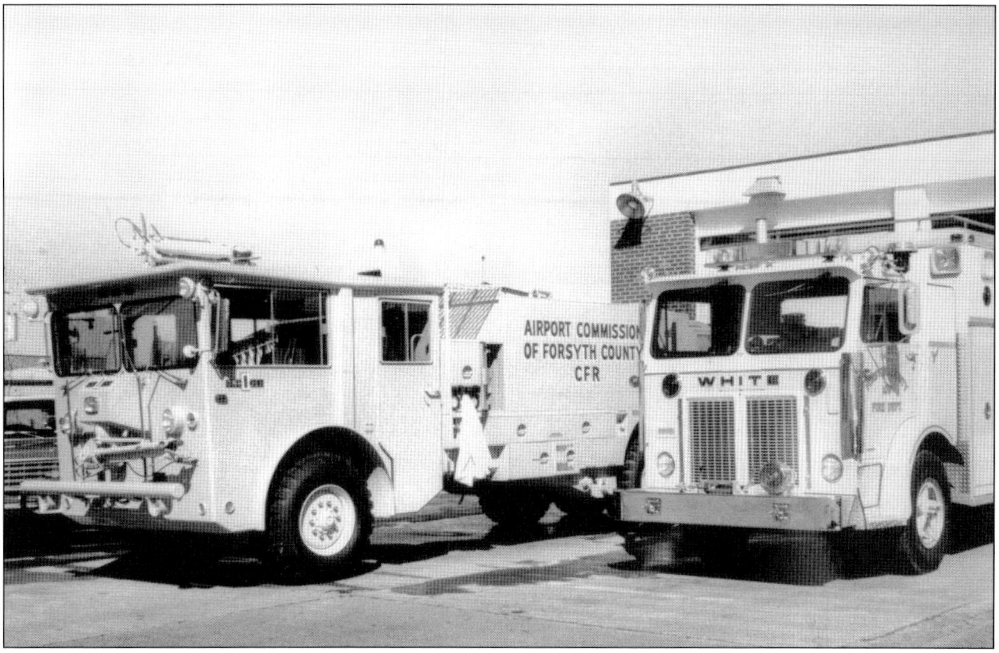

In 1975, the Forsyth County CFR truck was housed at the airport fire station along with the county's heavy rescue truck. (Courtesy of FCFHS.)

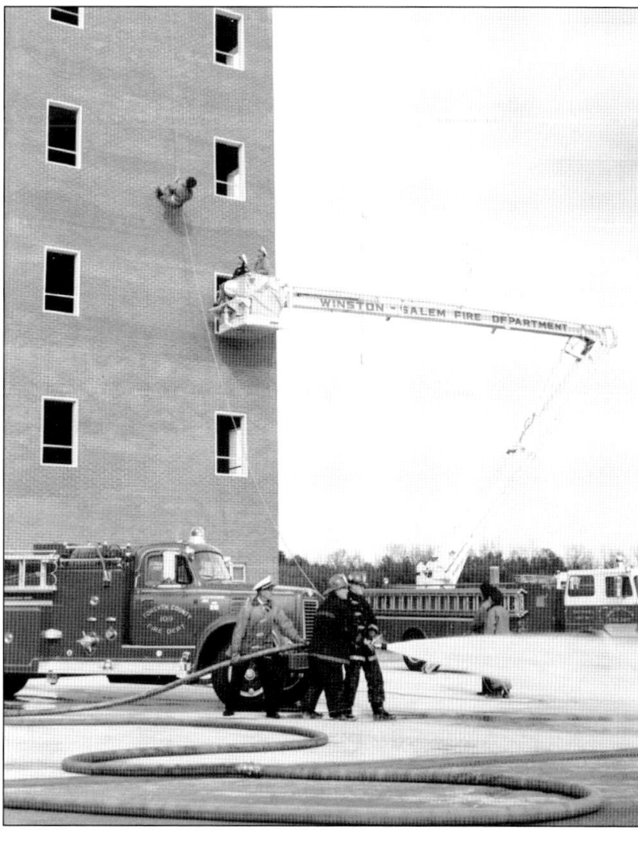

The Forsyth County 109 truck was used in training alongside the Winston-Salem Fire Department at the city training grounds. (Courtesy of FCFPL.)

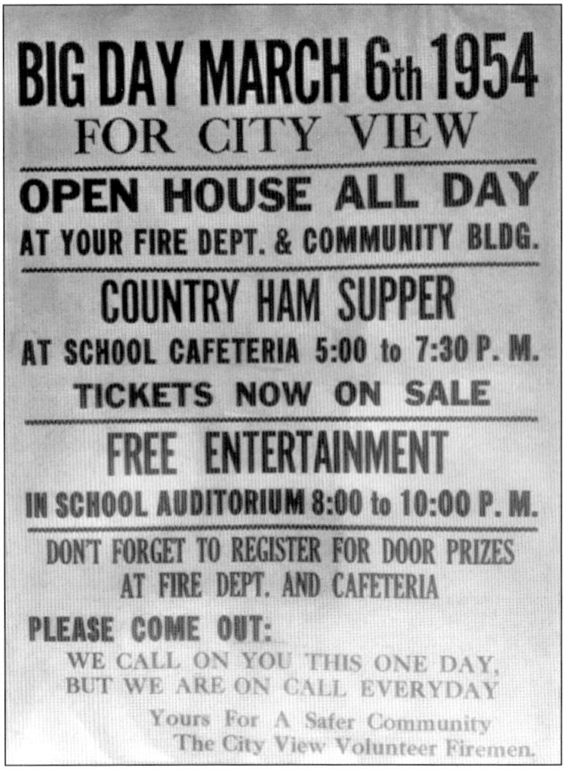

This is a poster advertising a country ham supper. This event was used as a fundraiser for the City View Volunteer Fire Department. Often, local businesses in the district would donate door prizes and food to help with fundraising efforts. Funds were needed to maintain the department's building and apparatus as well as purchase new equipment. (Courtesy of FCFHS.)

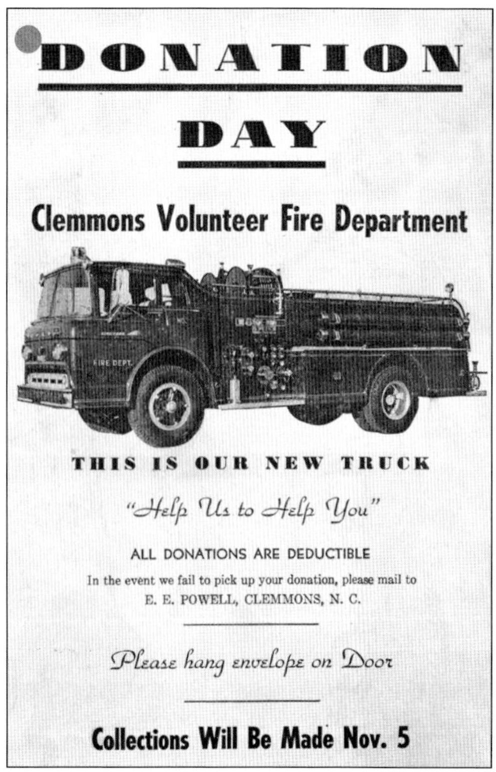

Firemen donation days were often a countywide effort. Department members would deliver envelopes door to door to raise funds for their department. As communities grew larger, better apparatus and new turnout gear for volunteer members was greatly needed. (Courtesy of FCFHS.)

The Forsyth County Volunteer Firemen's Association hosted an annual Firemen's Muster. All departments took the opportunity to show the town and other departments their skills. The muster ended with a parade and beauty contest. (Courtesy of FCFHS.)

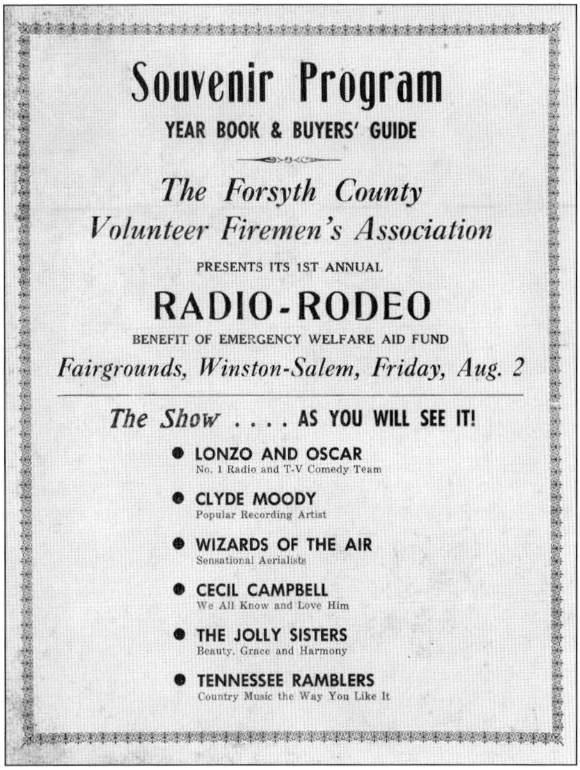

Souvenir Program
YEAR BOOK & BUYERS' GUIDE

The Forsyth County Volunteer Firemen's Association

PRESENTS ITS 1ST ANNUAL

RADIO-RODEO

BENEFIT OF EMERGENCY WELFARE AID FUND

Fairgrounds, Winston-Salem, Friday, Aug. 2

The Show . . . . AS YOU WILL SEE IT!

- LONZO AND OSCAR
  No. 1 Radio and T-V Comedy Team
- CLYDE MOODY
  Popular Recording Artist
- WIZARDS OF THE AIR
  Sensational Aerialists
- CECIL CAMPBELL
  We All Know and Love Him
- THE JOLLY SISTERS
  Beauty, Grace and Harmony
- TENNESSEE RAMBLERS
  Country Music the Way You Like It

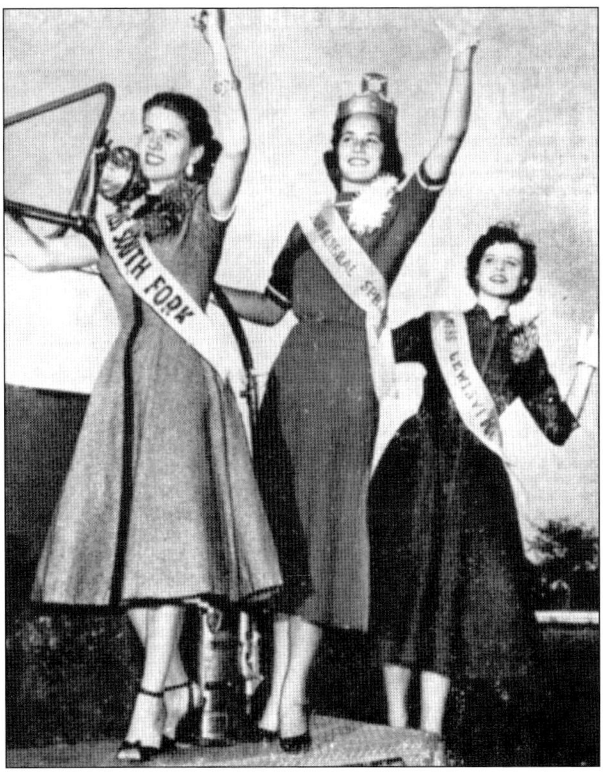

From left to right, beauty queens Ruby Hogue, Jean West, and Paige Jones wave to the crowd at the Firemen's Muster parade in 1955. (Courtesy of FCFHS.)

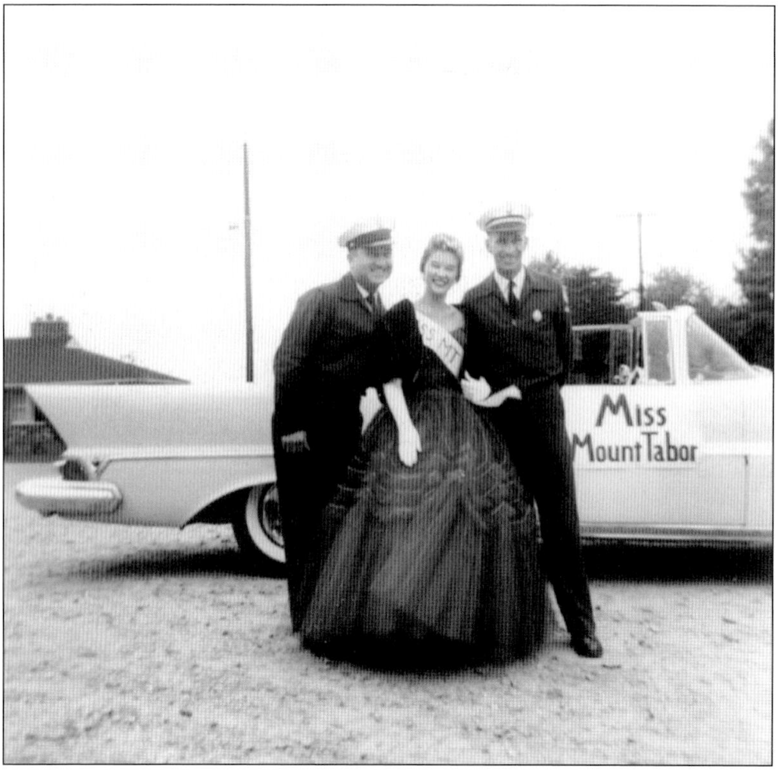

Miss Mount Tabor is escorted by two Mount Tabor volunteer members at the Firemen's Muster in 1955. (Courtesy of FCFHS.)

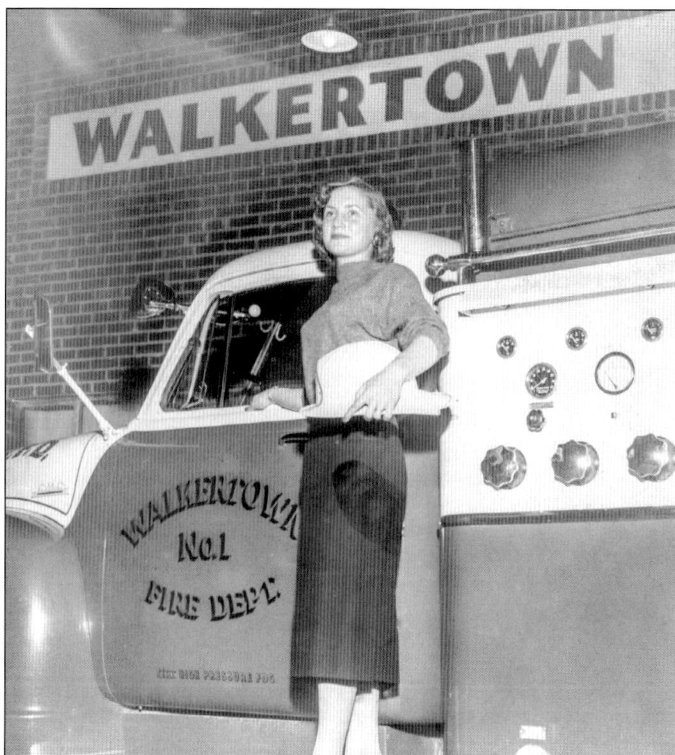

Miss Walkertown Jo Ann Hester was chosen to compete in the beauty event at the Firemen's Muster in 1955. (Courtesy of FCFHS.)

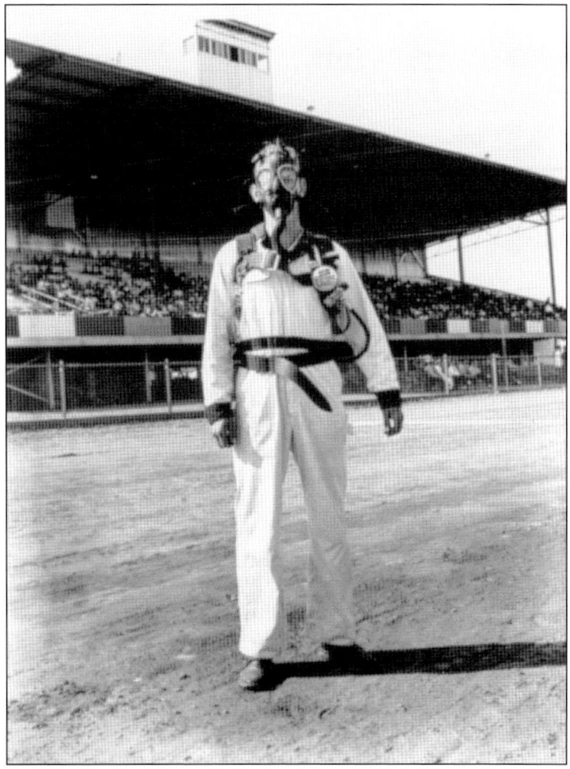

A Lewisville Volunteer Fire Department member is seen here performing a quick dress revolution at the Firemen's Muster. Volunteer members were expected to have all their gear on and be breathing off their air in under a minute. This became increasingly more difficult as many departments did not have enough gear for all their members. (Courtesy of FCFHS.)

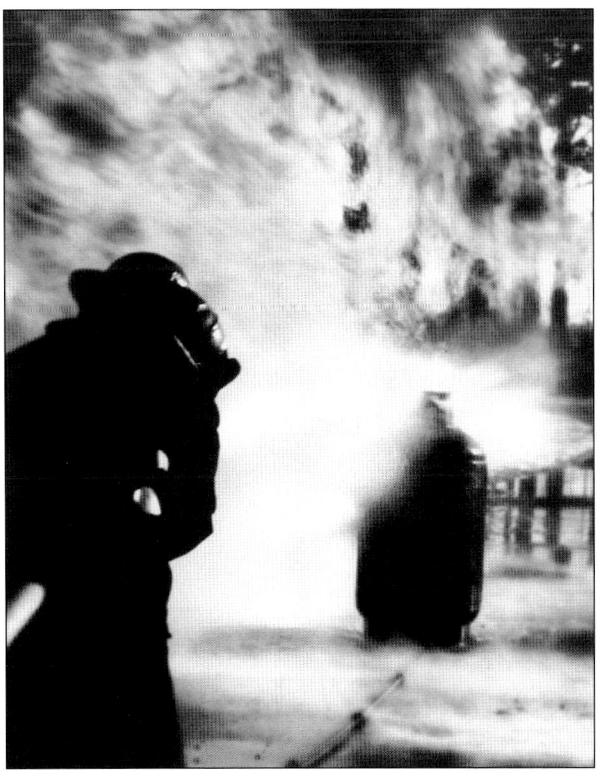

A volunteer firefighter is performing a hose advancement on a propane tank at the Firemen's Muster. (Courtesy of FCFHS.)

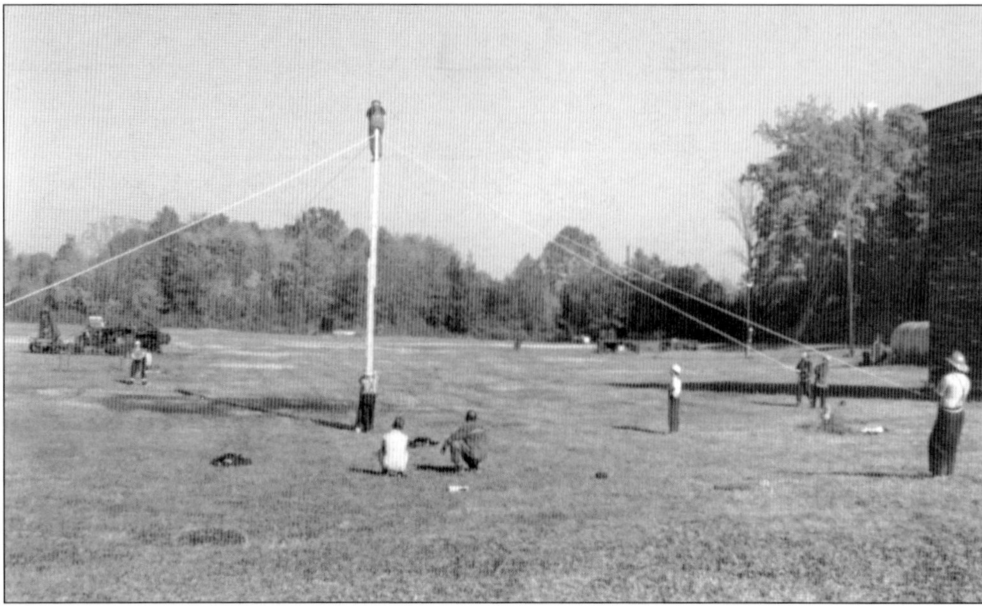

The City View Volunteer Fire Department shows how to set up a ladder as a derrick at the Firemen's Muster. Several skills were performed at the Firemen's Muster, and the volunteer firefighters rarely disappointed on practical tasks. (Courtesy of FCFHS.)

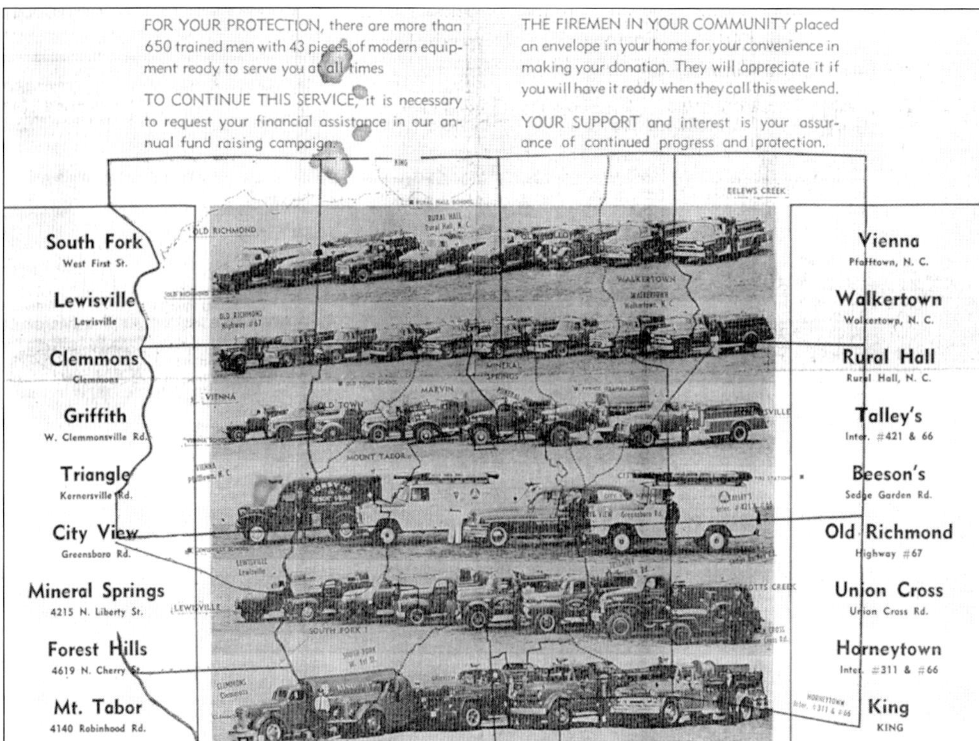

The Firemen's Muster was an opportunity for all fire and rescue departments to gather together, not only to show their skills, but also to learn from each other and become better firefighters. (Courtesy of FCFHS.)

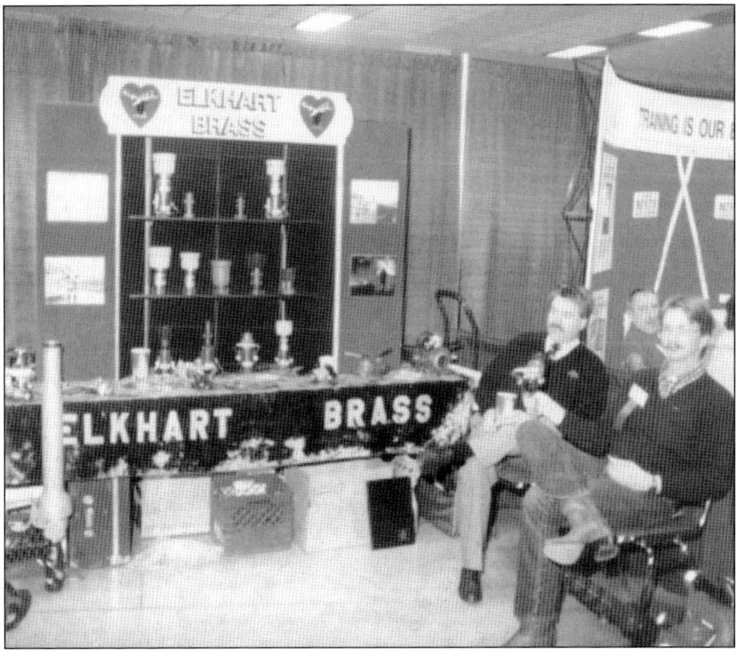

A fire safety seminar was hosted in Forsyth County to promote fire safety and prevention among schools, churches, industry, insurance groups, and fire departments. Pictured here is a vendor, Elkhart Brass, displaying firefighting equipment. Seminars allowed departments the opportunity to connect directly to businesses like Elkhart Brass. (Courtesy of FCFHS.)

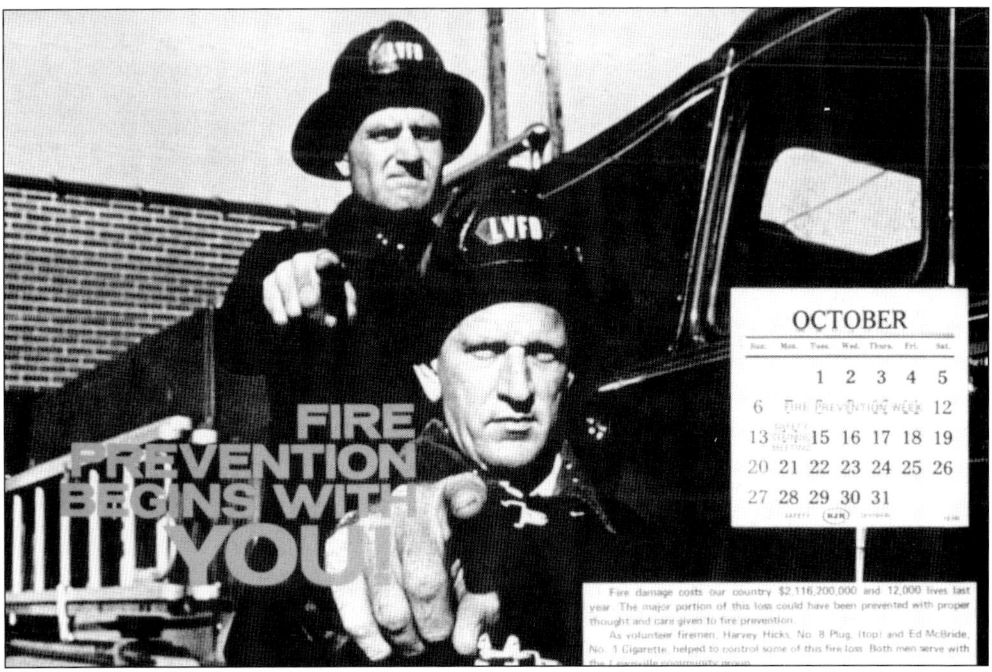

Every fire department in Forsyth County knew the benefits of public education. When the public was taught fire safety and prevention, the job of protecting the community was much easier. Pictured is a public education poster featuring Lewisville Volunteer Fire Department members Harvey Hicks (left) and Ed McBride in 1968. (Courtesy of FCFHS.)

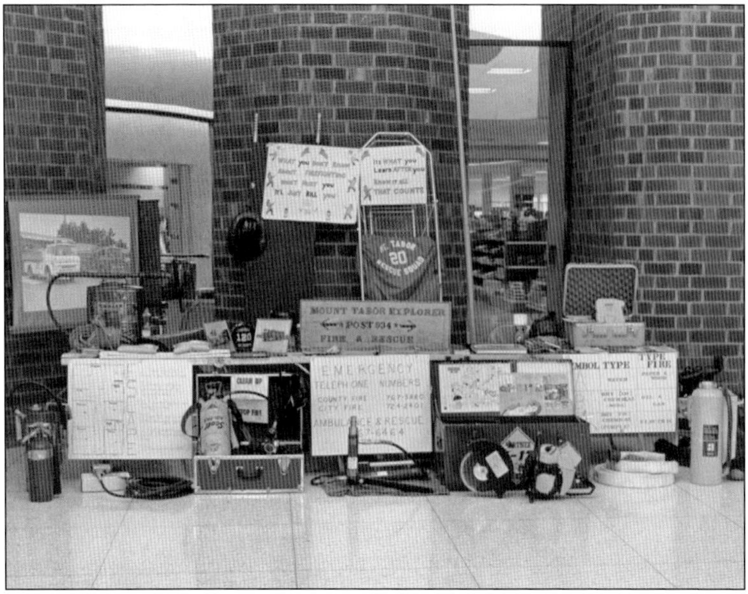

Most fire and rescue departments used public events to display equipment, educate on fire safety, and recruit volunteers. Pictured is the Mount Tabor Volunteer Fire Department displaying its equipment at a public education event at a high school. (Courtesy of FCFHS.)

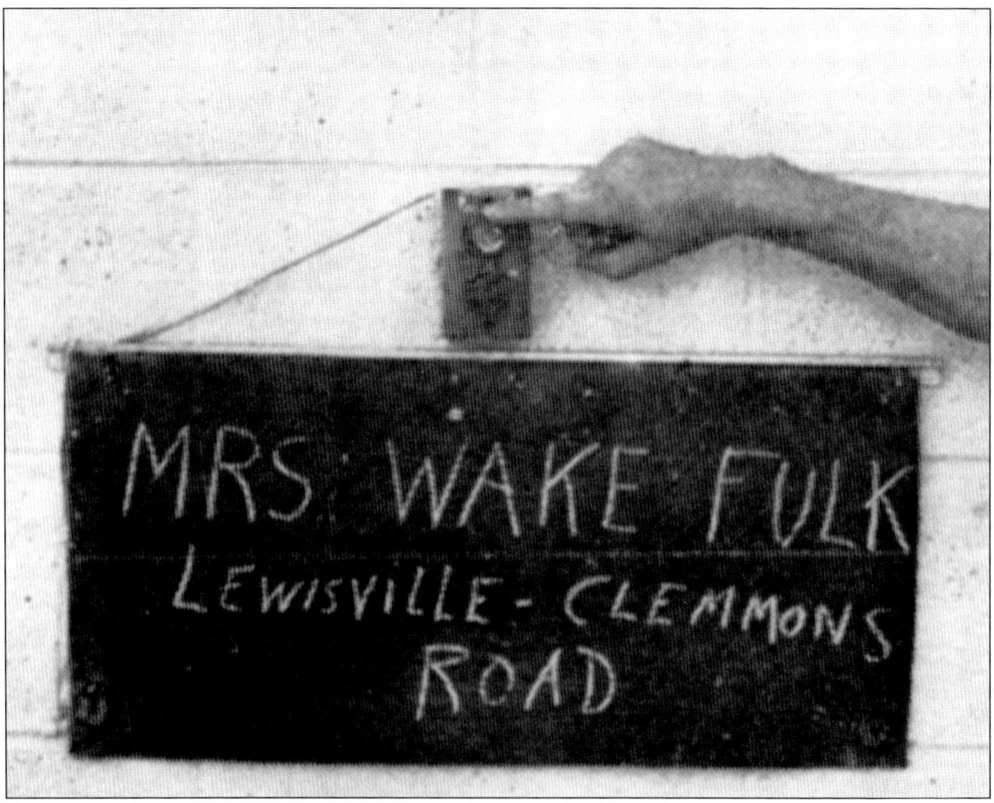

Before central communications, community members in rural areas would use a chalkboard like this one at the Lewisville Volunteer Fire Department station. Community members would write where the event was with chalk and push a button to notify the department. The first to respond to the department would drive the apparatus and throw small bags of limes out the window of the apparatus at turns to show the way to the fire. (Courtesy of FCFHS.)

Each fire department gave out stickers (like Mineral Springs, shown here) with different phone numbers to call in case of a fire. The phone numbers typically called individual volunteer firefighters at their homes. If one number did not answer, the next number on the list was called. This was the system before 911 was a countywide service. (Courtesy of FCFHS.)

Courtesy
INSURANCE SERVICE CO.
Fire and Auto — Phone PA 2-7187

Phone:	Day	Night
MINERAL SPRINGS VOLUNTEER FIRE DEPT., INC.	PA 4-2026	1st Call PA 3-6097 2nd Call PA 5-2631 3rd Call PA 5-5411 4th Call PA 4-9478
COUNTY FIRE DEPT.	PA 4-1770	PA 4-1770

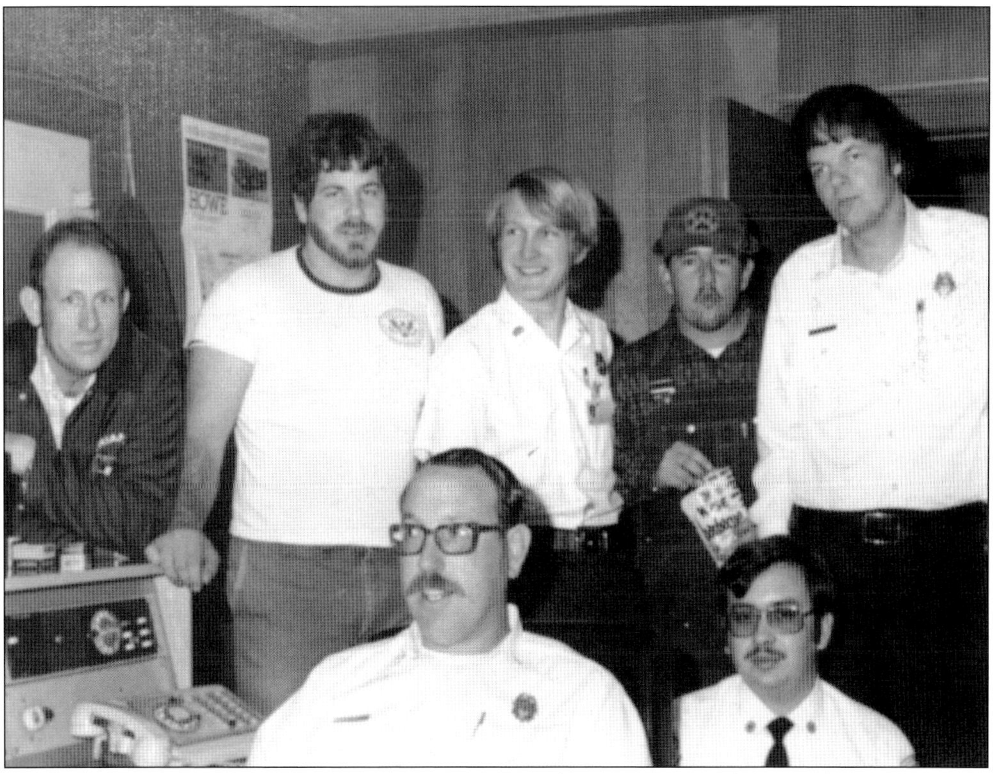

Pictured are some of the first radio operators in Forsyth County. From left to right are (first row) Denis Whicker and Garry Blankenship; (second row) Ray Smith, Jim Southern, Ray Deans, Rickey Weavil, and Rick Plunkett. The Forsyth County Fire Department radio system was the first rural radio communications division in the Southeast. (Courtesy of FCFHS.)

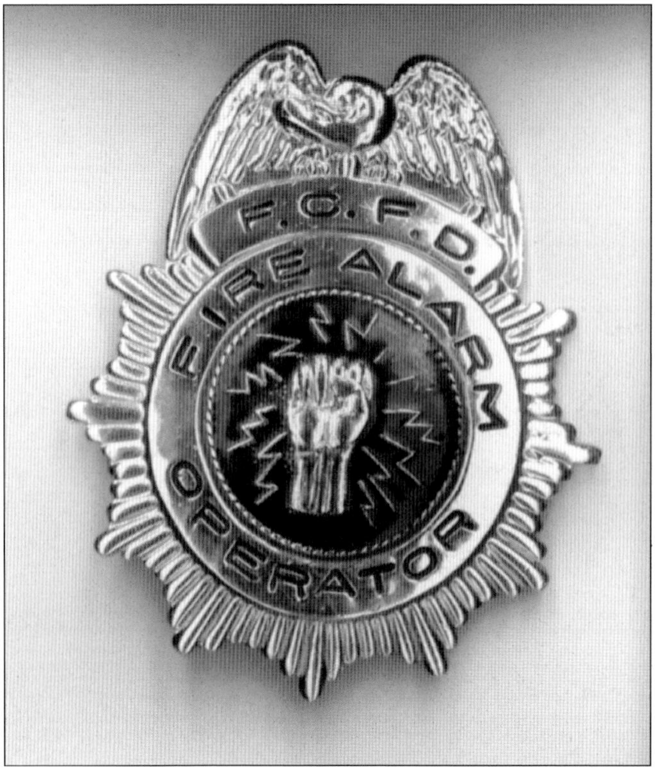

Central Communications was established in 1964 through funds from Volunteer Firemen's Association members and a Department of Civil Defense grant. Pictured here is the Forsyth County telecommunications badge from 1965. (Courtesy of FCFHS.)

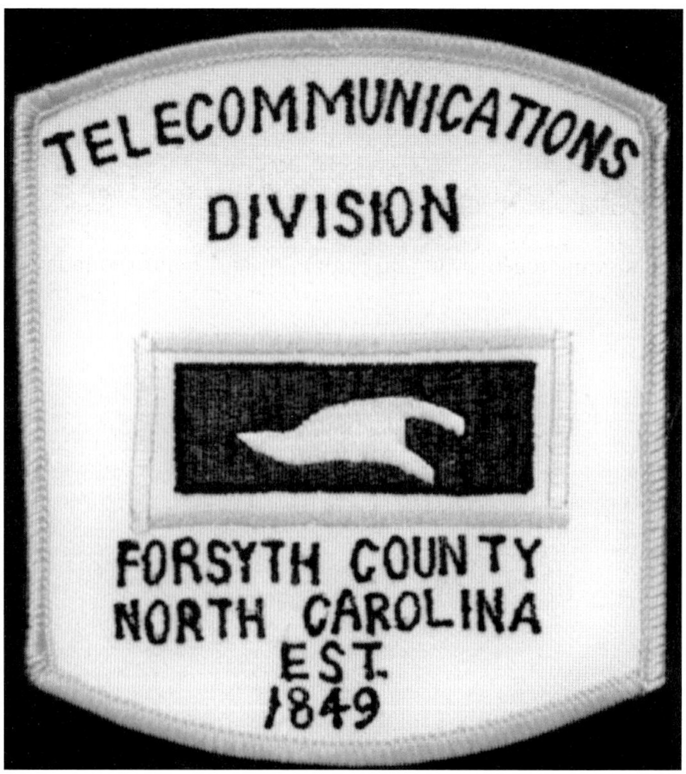

Pictured is the Forsyth County telecommunications division patch. The patch featured Forsyth County's year of establishment and logo. (Courtesy of FCFHS.)

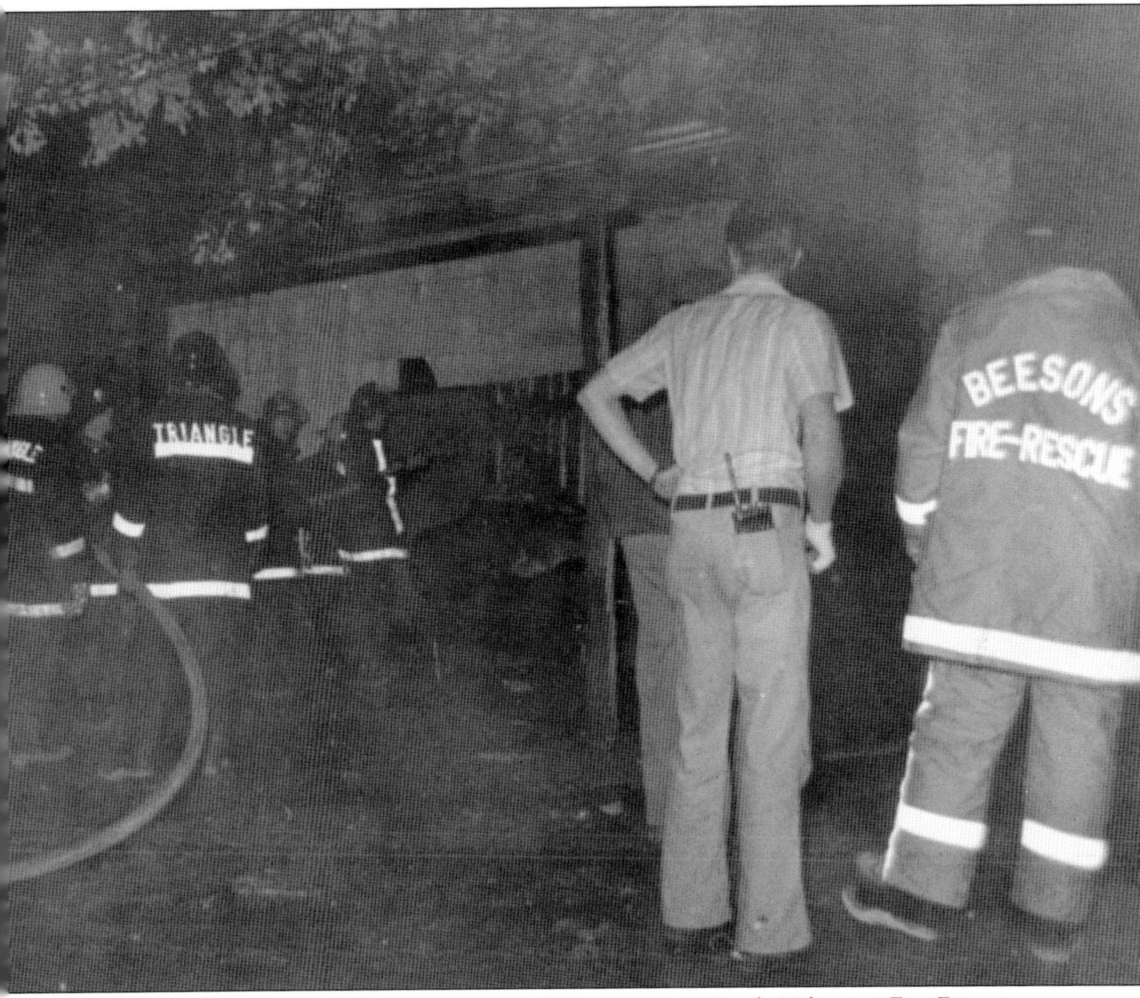

The Triangle Volunteer Fire Department and Beeson Cross Roads Volunteer Fire Department are working together on a fire scene. A large handheld radio can be seen in the pocket of the on-scene command. Radios allowed for easier communication on scene. (Courtesy of FCFHS.)

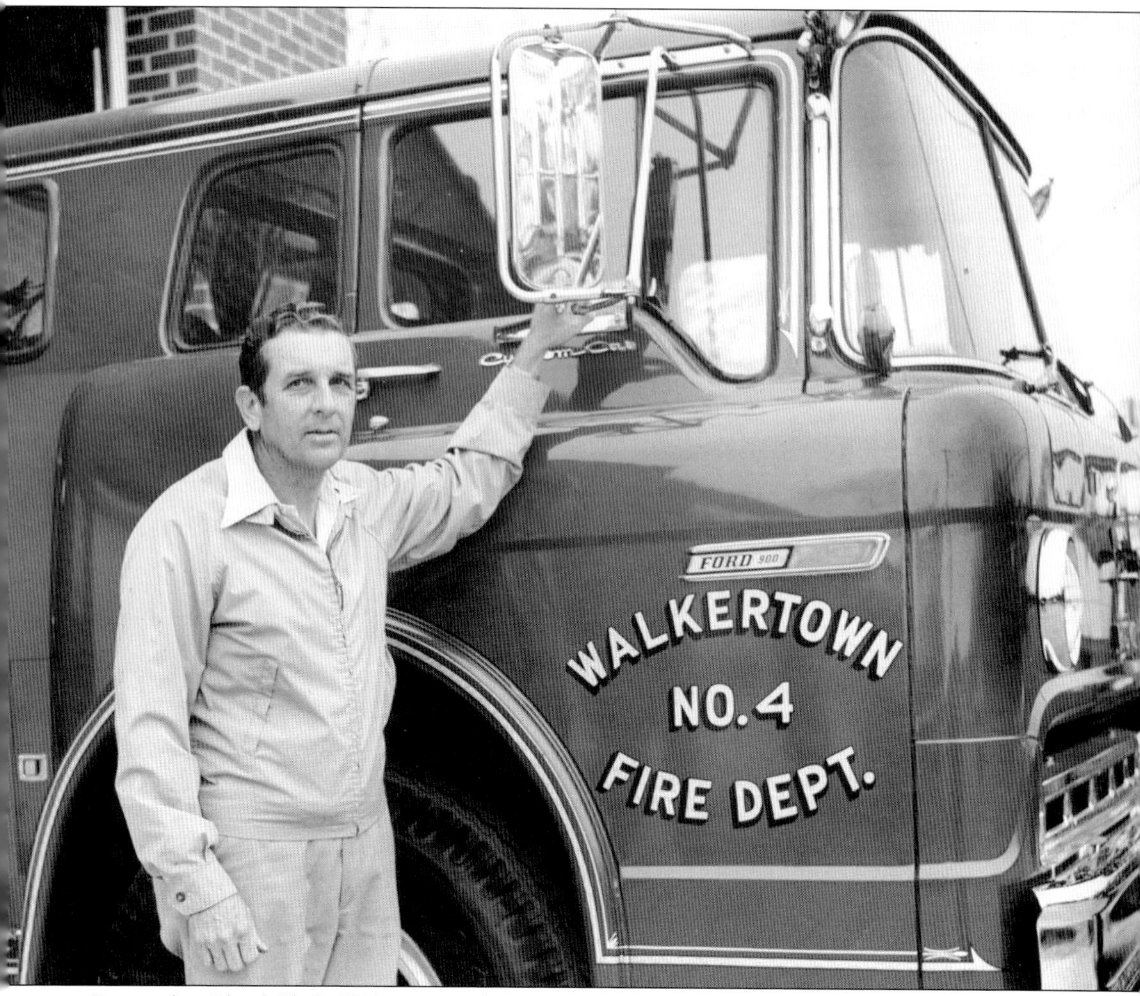

Pictured is Chief Clyde Webster standing next to the Walkertown Volunteer Fire Department truck. Webster served as fire chief from 1953 to 1978. (Courtesy of FCFHS.)

Four

RESCUE AND EMS

The history of rescue squads in Forsyth County is one of the oldest in the nation. The rescue needs of Forsyth County were handled by squads established by local civic clubs. The need became so great that the fire service trained to not only serve in a fire protection role, but many of the departments have also now added a rescue component. Whenever asked to change to serve the public, the emergency responder has always boldly stood up and gone to work. The rescue model was established, and patient transport functions were added to supplement funeral homes, which at the time were the primary transport for the sick and injured.

The Forsyth County Ambulance Service was established by the Forsyth County Board of Commissioners in 1968. B.W. Atkinson, a captain from the Forsyth County Fire Department, was appointed director. Ambulances were dispatched by the Forsyth County Fire Department. The Forsyth County Emergency Medical Services later grew to provide advanced life support services. This paved the way to training the local fire departments in medical care. Many of the first firefighters in Forsyth County became leaders in emergency medical services. Just as it has since 1968, Forsyth County Emergency Medical Services has always been a leader in the emergency medical services industry. These men and women provide the most distinguished prehospital emergency medical care available in a compassionate manner.

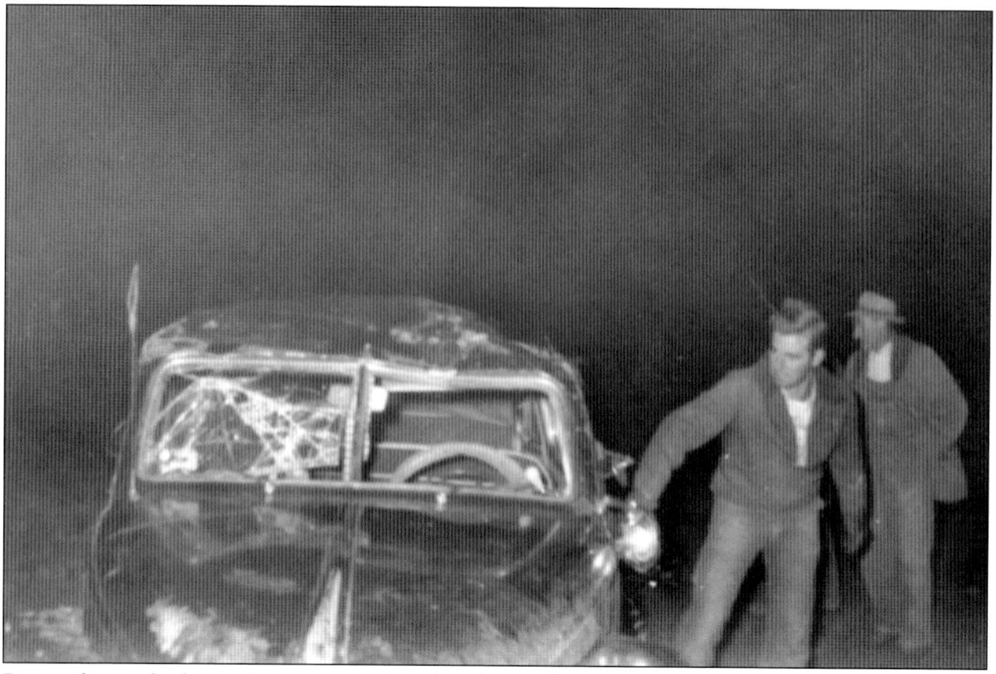

Pictured is a vehicle accident on a rural road in the Walkertown area. Accidents like this happened more as the population grew. (Courtesy of FCFHS.)

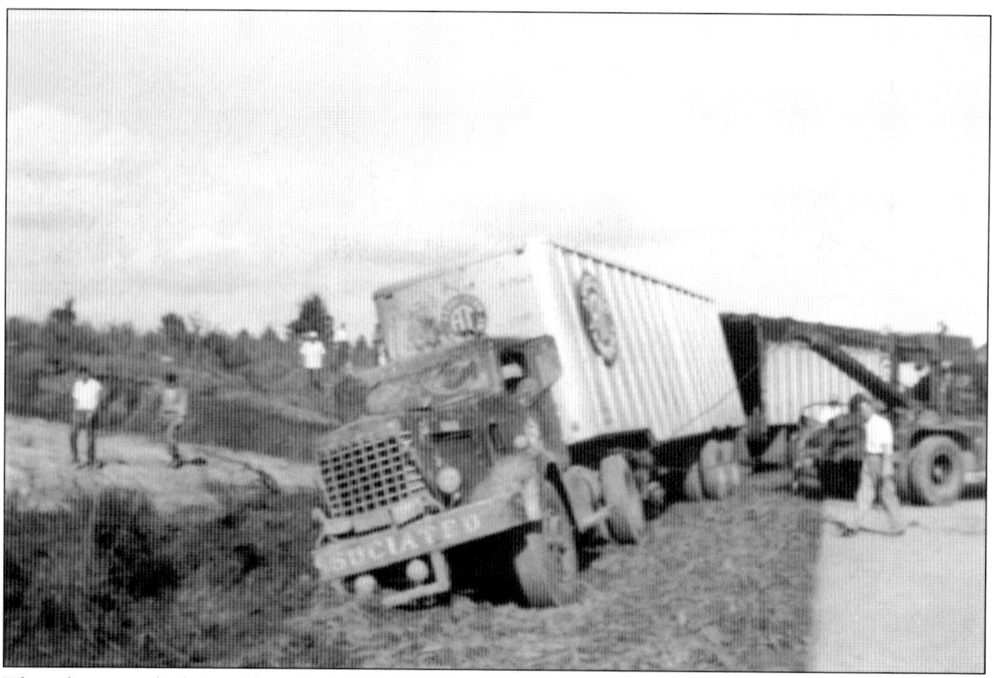

This photograph shows a large Associated Transport truck accident on rural Highway 66 in Forsyth County in 1956. The need was growing to provide more specialty rescue equipment throughout the county. (Courtesy of FCFHS.)

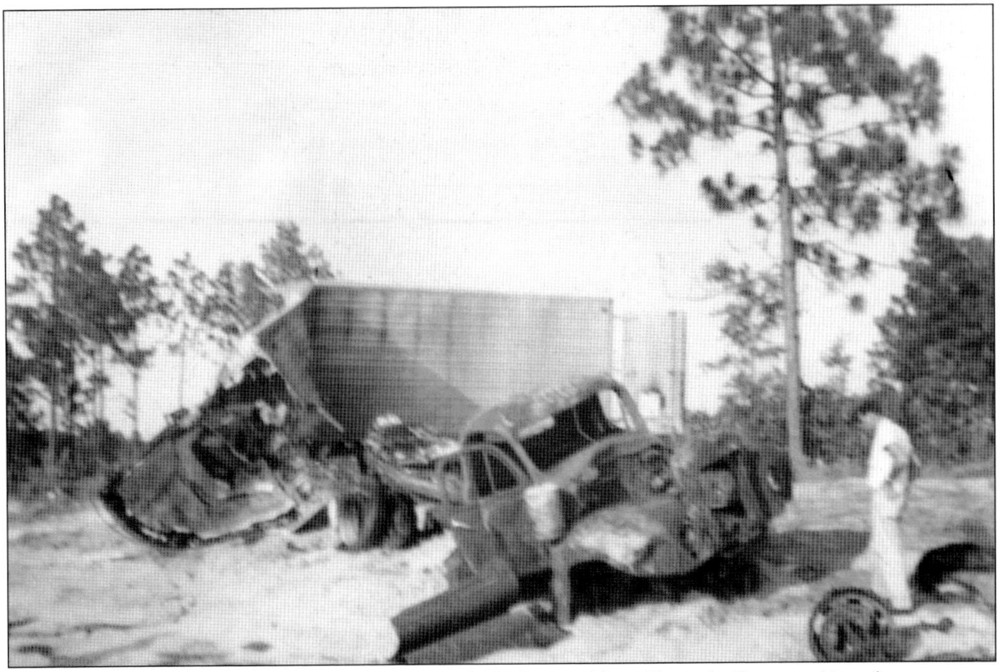

This is another large truck accident on rural Highway 66 in Forsyth County. Fire departments would respond to incidents like this, but the tools needed for appropriate rescue measures were not always available in time. (Courtesy of FCFHS.)

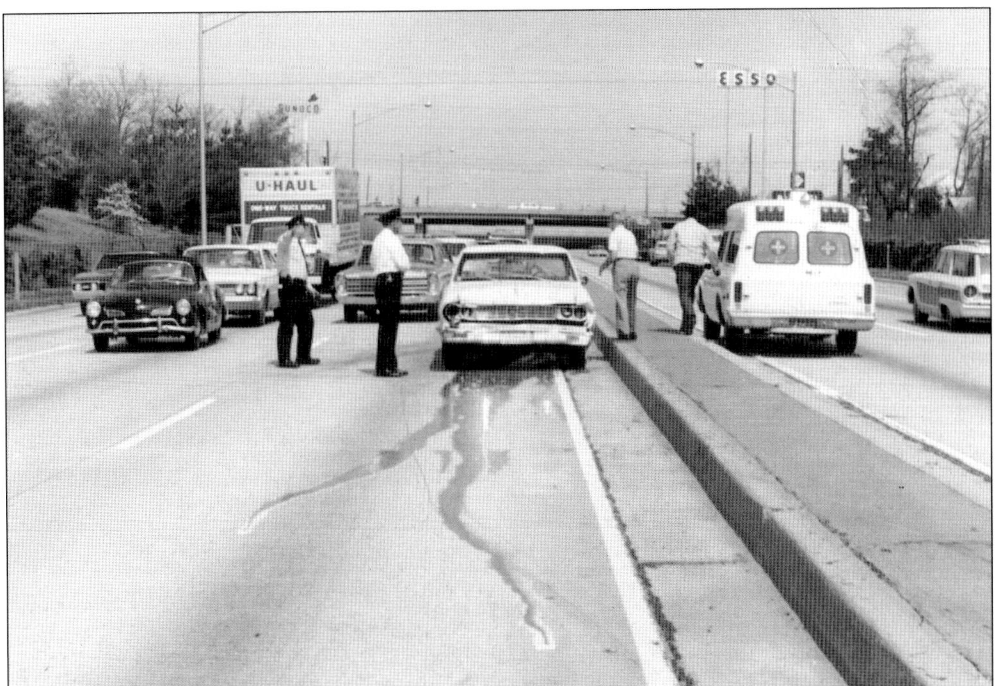

This was an accident on a highway in Winston-Salem near Fourth Street in 1975. The need for rescue services was felt both in the city and the county. (Courtesy of FCFHS.)

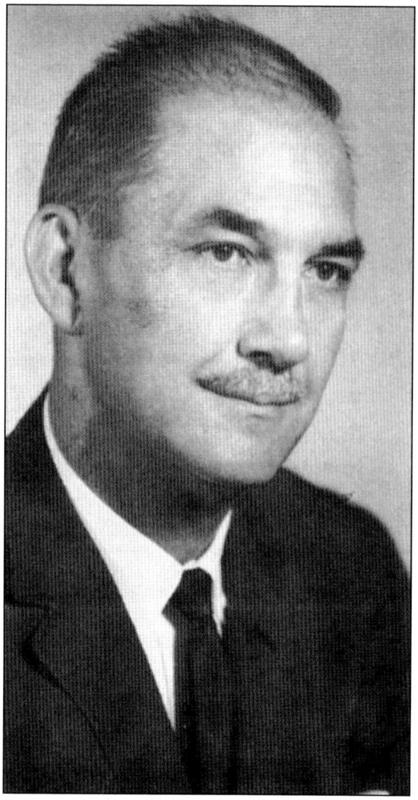

In 1937, Roy Kane (pictured) and Thurmond Chatham founded the volunteer Winston-Salem Rescue Squad. This was the first such agency in North Carolina and one of the first in the country. (Courtesy of FCFHS.)

The Winston-Salem Rescue Squad was a source of durable medical equipment, such as oxygen tanks, hospital beds, and even iron lungs for polio treatment. The rescue squad responded to auto and boating incidents. (Courtesy of FCFHS.)

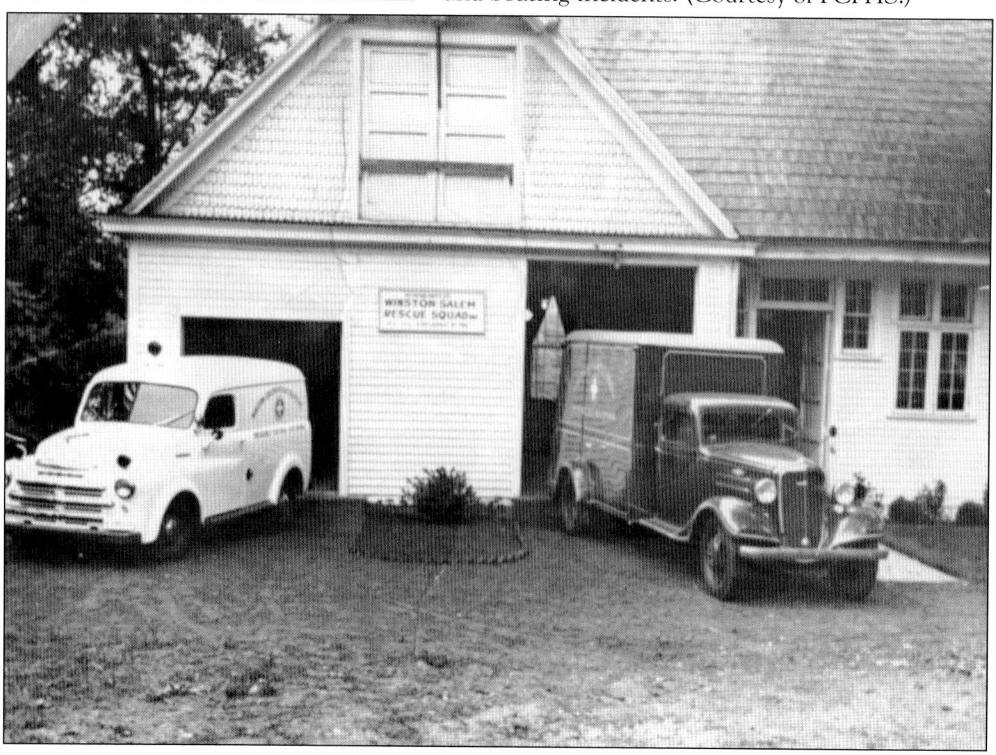

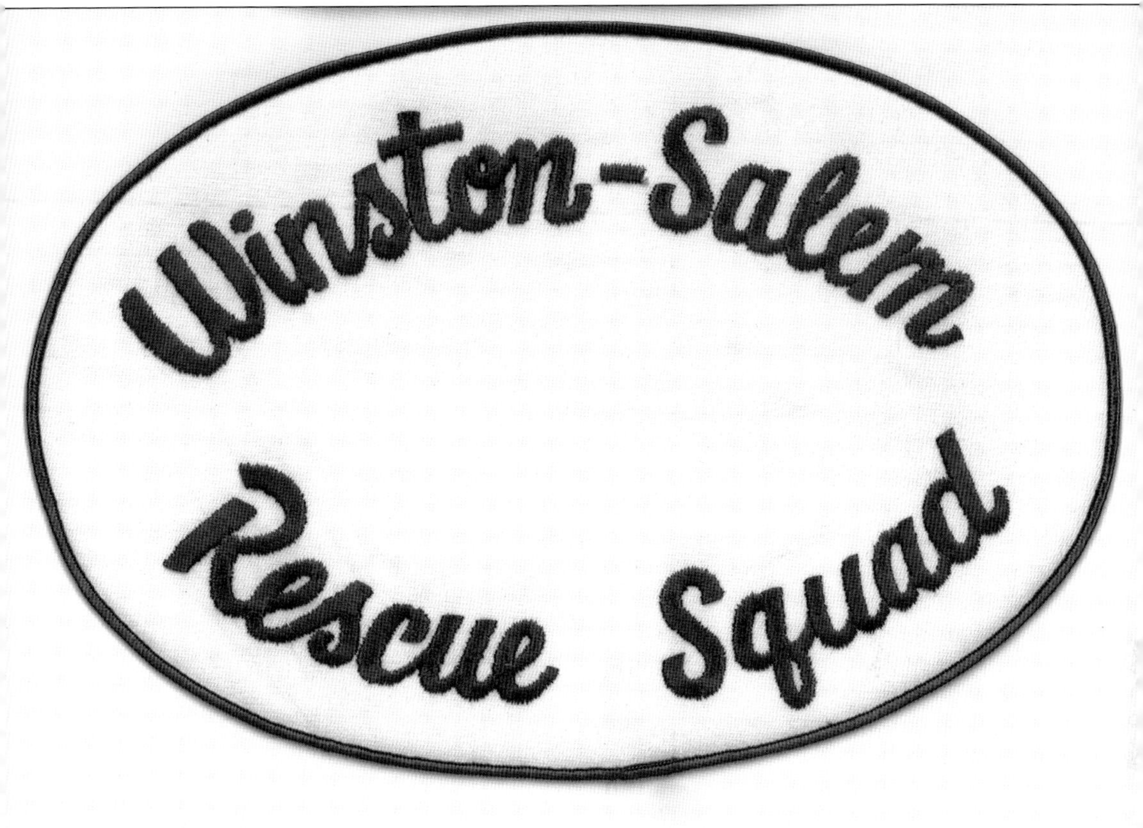

This is the Winston-Salem Rescue Squad patch that was worn on the back of the members' coveralls. (Courtesy of FCFHS.)

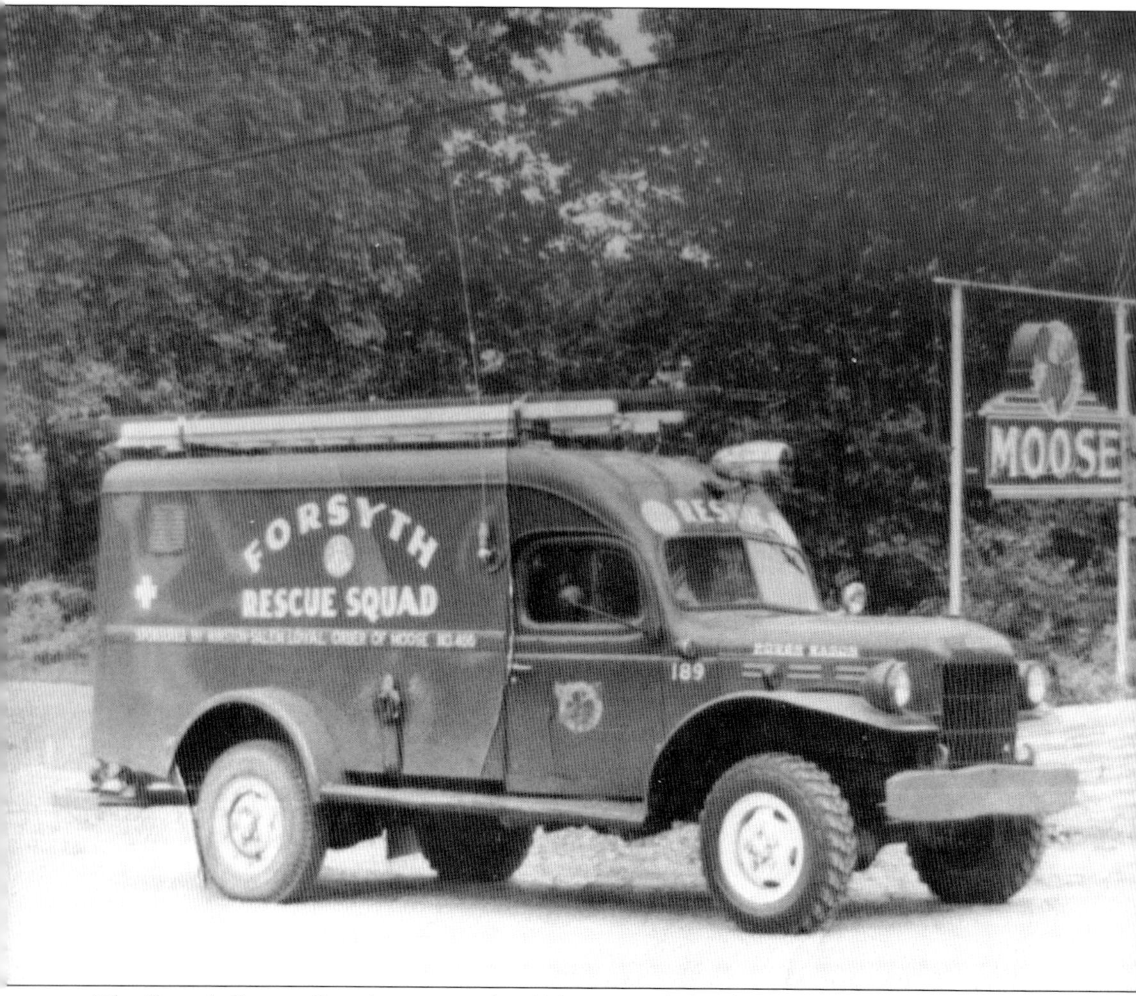

The Forsyth Rescue Squad was started in February 1959 by the North Carolina Civil Defense Office. Its mission was to engage in rescue and first aid activities for the public's welfare. The rescue squad was housed and sponsored by the Winston-Salem Order of the Moose No. 466. Pictured is the original 1951 Dodge military-type ambulance. In 1978, along with the Winston-Salem Rescue Squad, an infant transport program was initiated for the North Carolina Baptist Hospital intensive care nursery. This allowed medical providers to treat and stabilize ill infants to be transported. (Courtesy of FCFHS.)

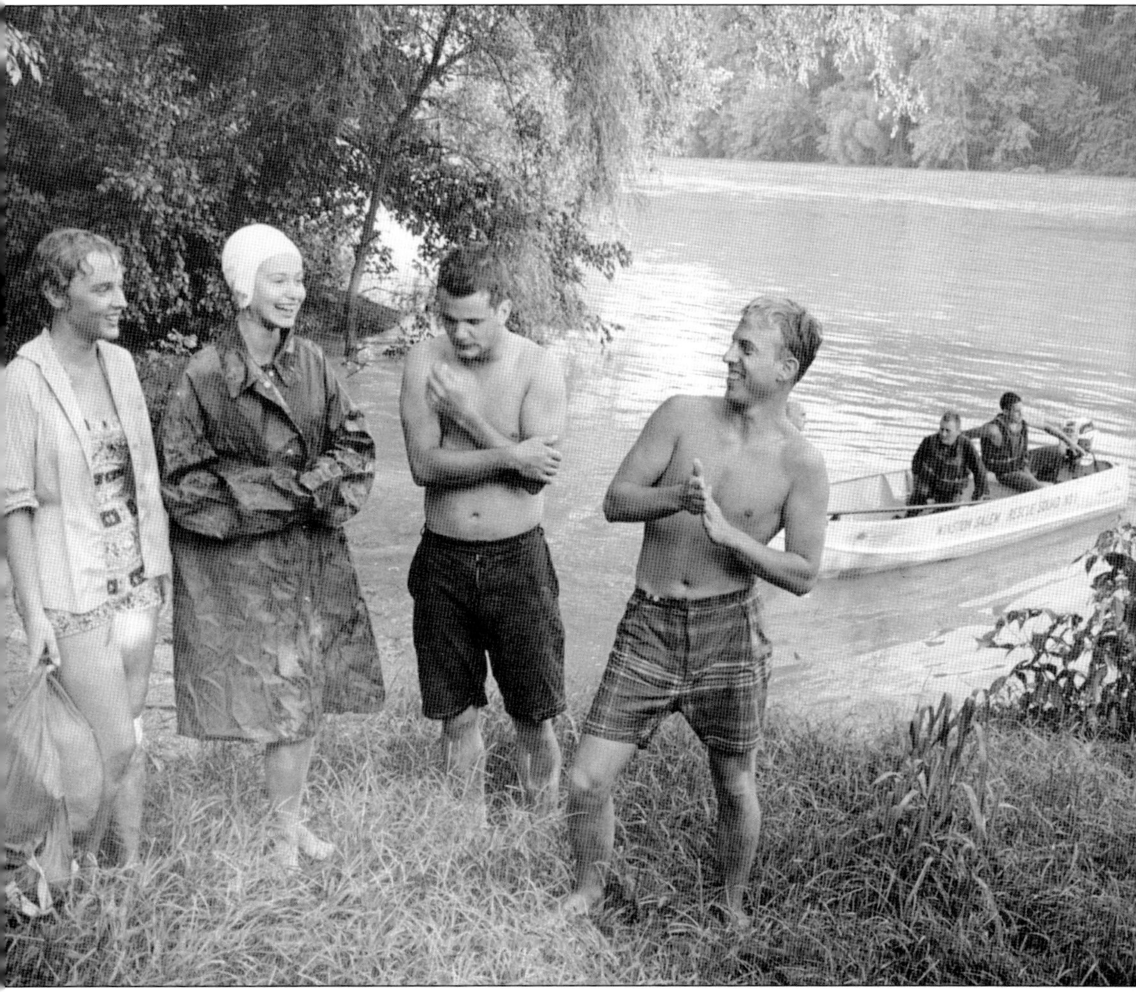

The Winston-Salem Rescue Squad is seen here on the Yadkin River in 1960, when boaters were involved in an accident near Idol's Dam. From left to right are Letitia Rice, Ann Herring (Miss North Carolina), Archer Glenn, and James Harrison. The Winston-Salem Rescue Squad boat is in the background. (Courtesy of FCPL.)

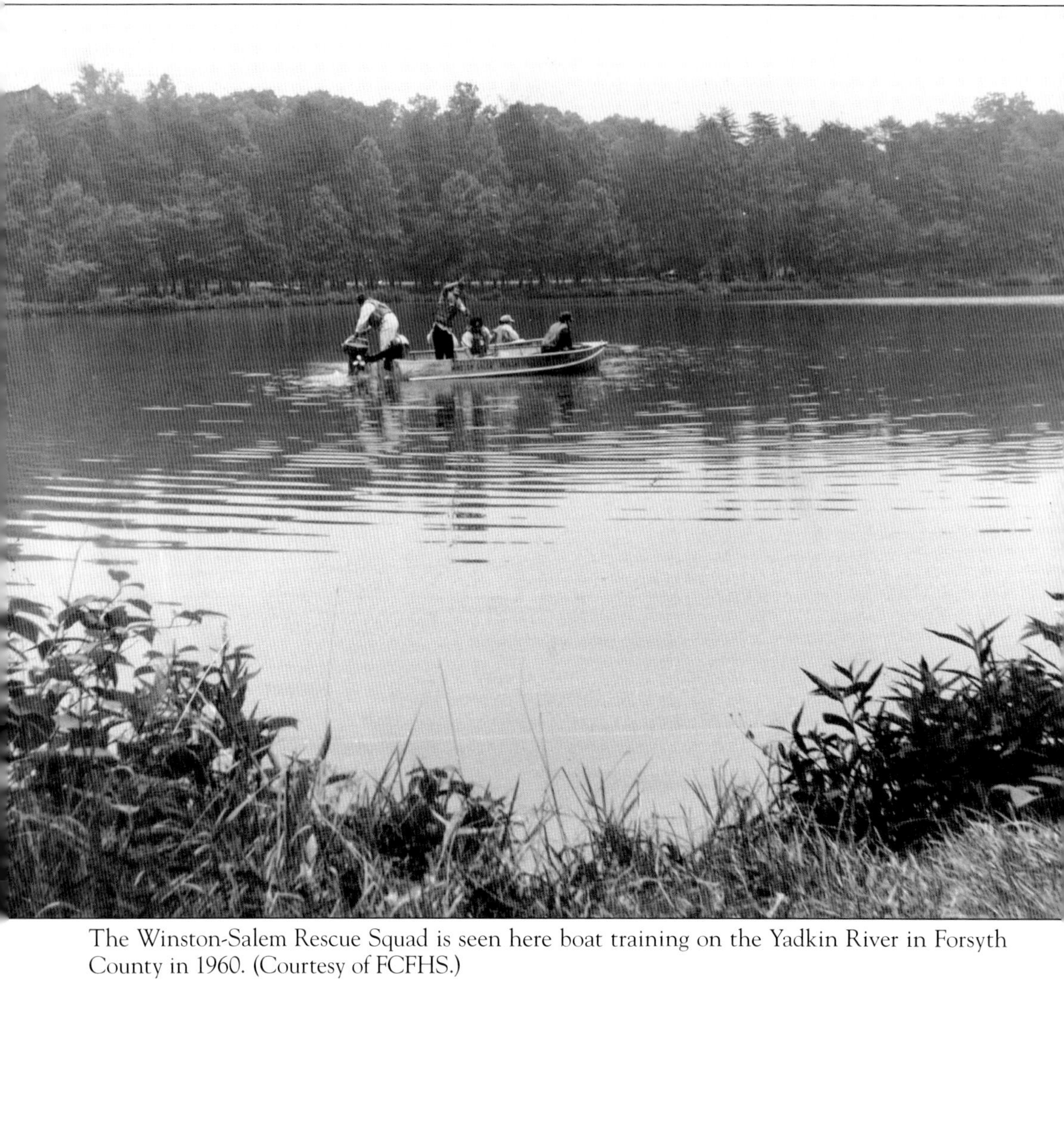

The Winston-Salem Rescue Squad is seen here boat training on the Yadkin River in Forsyth County in 1960. (Courtesy of FCFHS.)

The Winston-Salem and Forsyth Rescue Squads often worked and trained with the local fire departments. Pictured are squad members performing vehicle extrication training with fire department personnel. The Forsyth Rescue Squad was the first agency to acquire a Hurst tool, which allowed for much easier access for vehicle and machinery rescues. (Courtesy of FCFHS.)

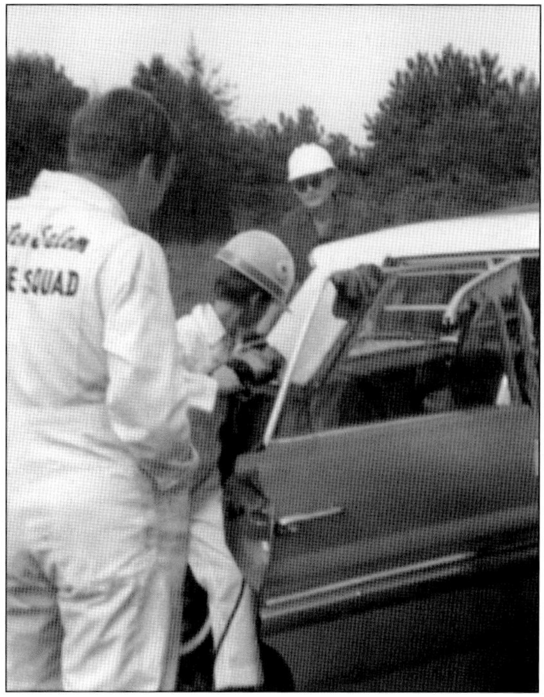

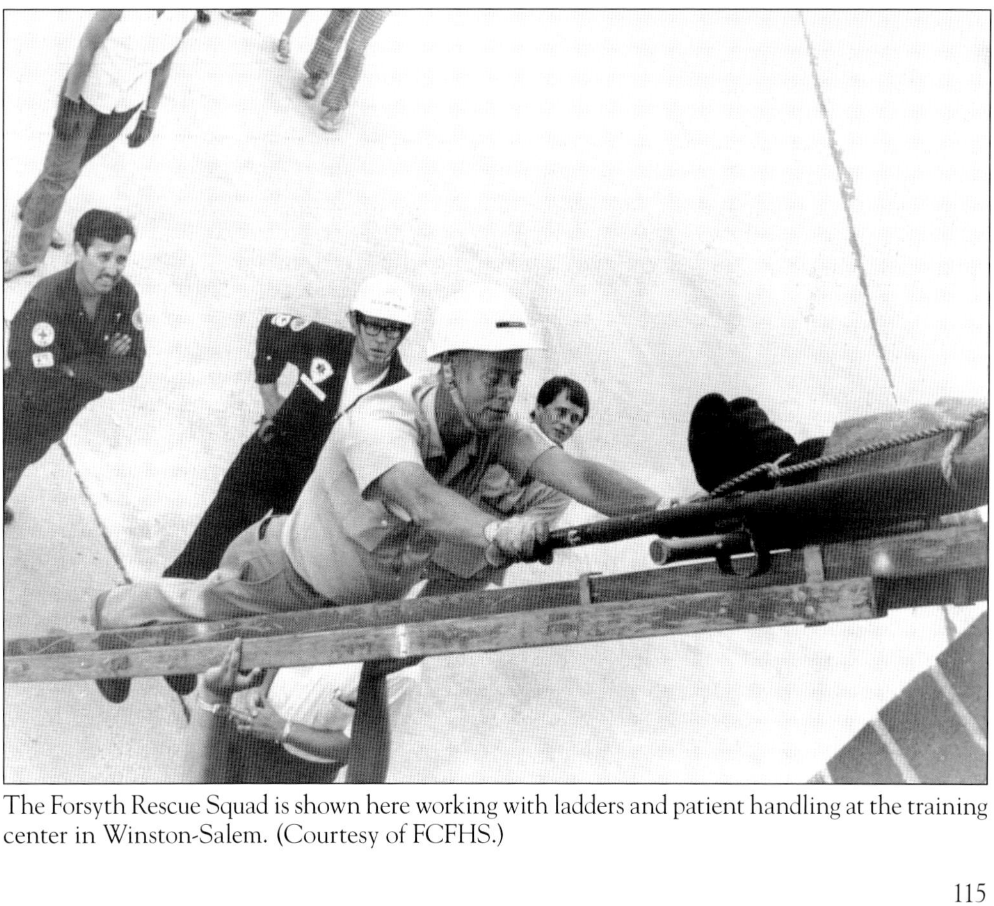

The Forsyth Rescue Squad is shown here working with ladders and patient handling at the training center in Winston-Salem. (Courtesy of FCFHS.)

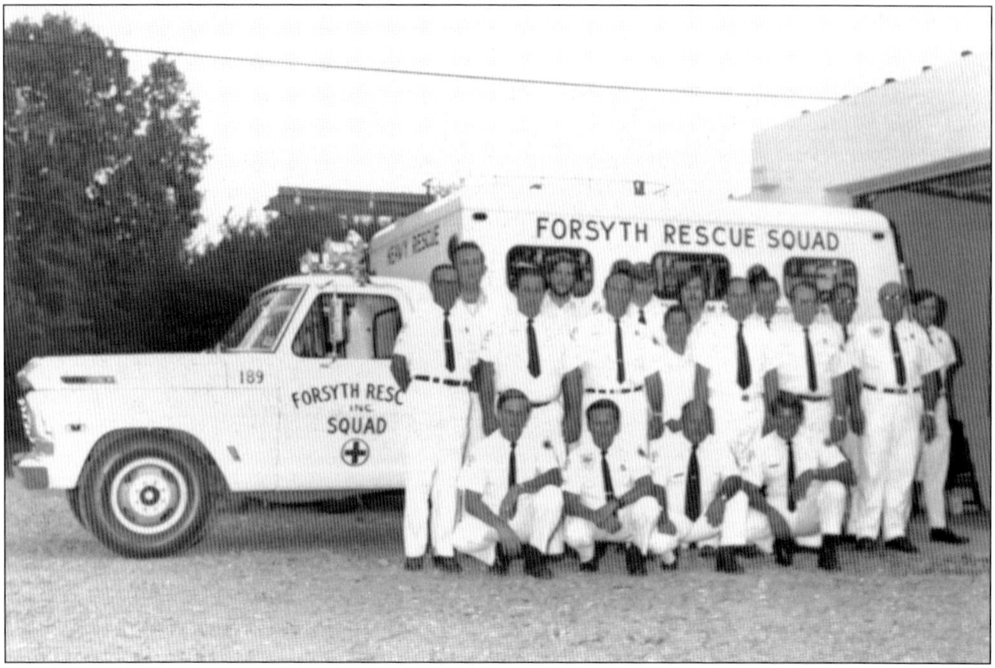

Pictured are members of the Forsyth Rescue Squad. In April 1973, the squad placed a modular-type ambulance in service. This was the first transport unit of this type in Forsyth County. The squad also began participating in the countywide ambulance assist program, which was a forerunner of today's first responder program. (Courtesy of FCFHS.)

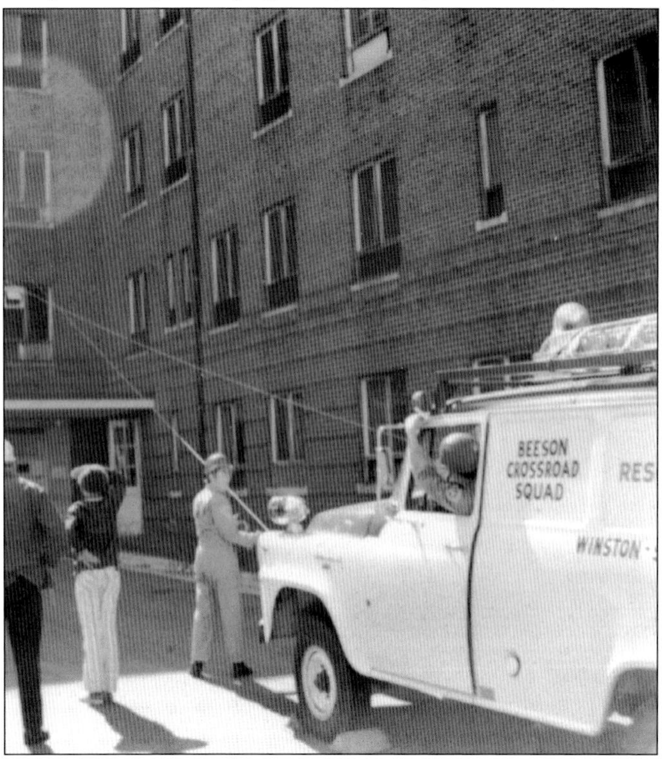

The Beeson Cross Road Volunteer Fire Department added rescue to its name in 1960. Its members are seen here assisting on a high-angle rescue call with the Forsyth Rescue Squad. (Courtesy of FCFHS.)

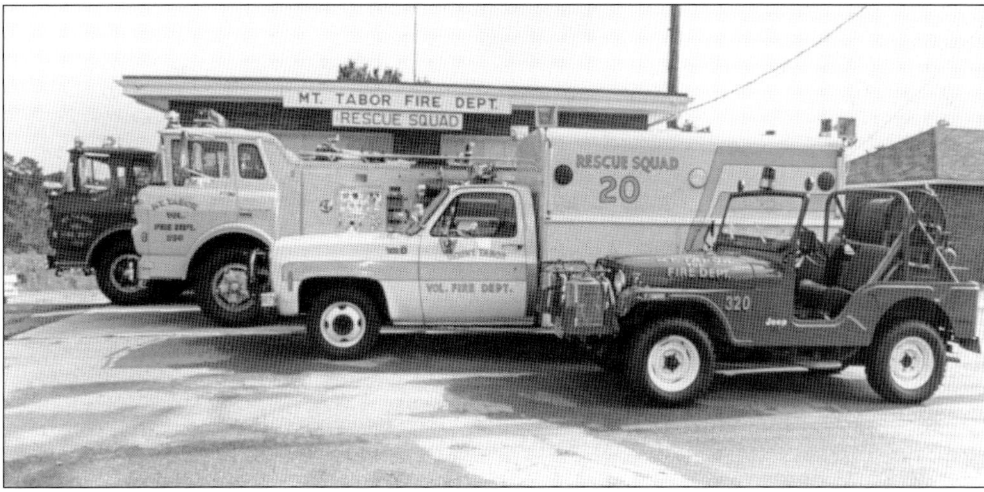

The Mount Tabor Volunteer Fire Department added a rescue squad due to the rise of rescue calls in 1976. (Courtesy of FCFHS.)

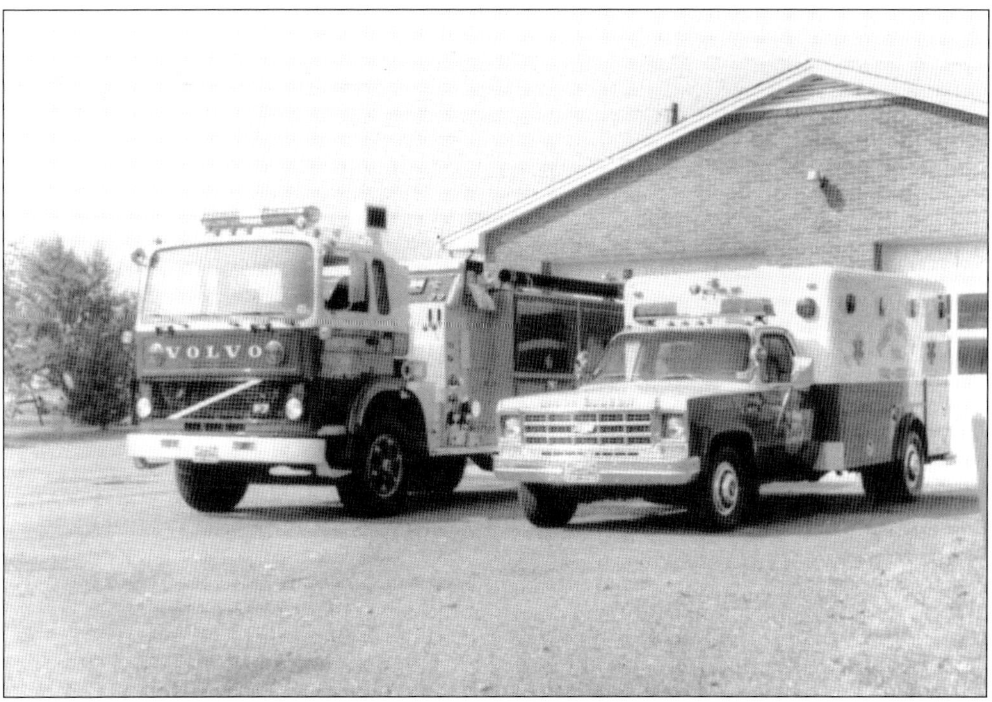

The Piney Grove Volunteer Fire Department began responding to rescue calls in 1973. (Courtesy of FCFHS.)

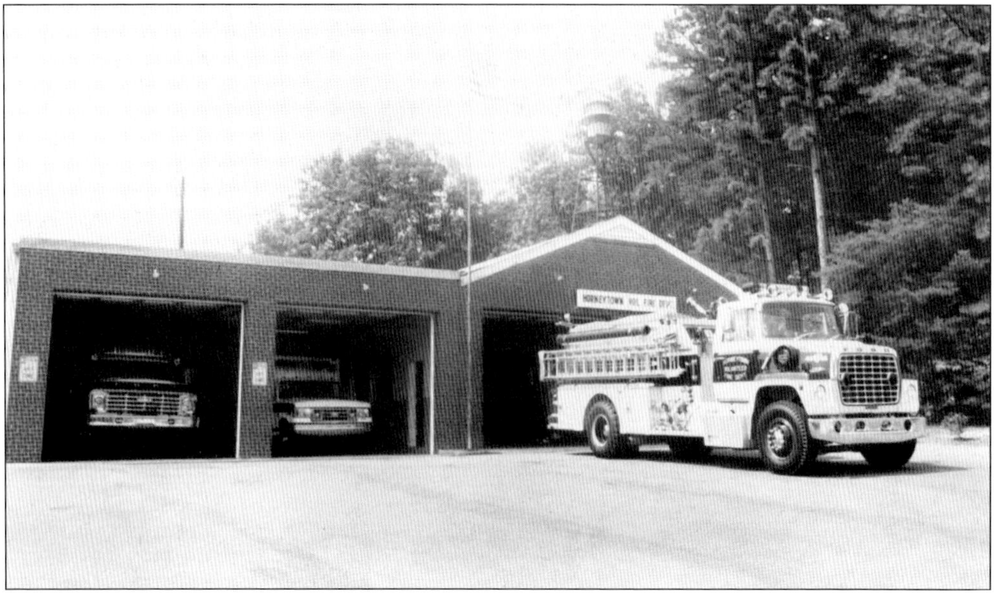

The Horneytown Volunteer Fire Department began responding to rescue calls in 1970. (Courtesy of FCFHS.)

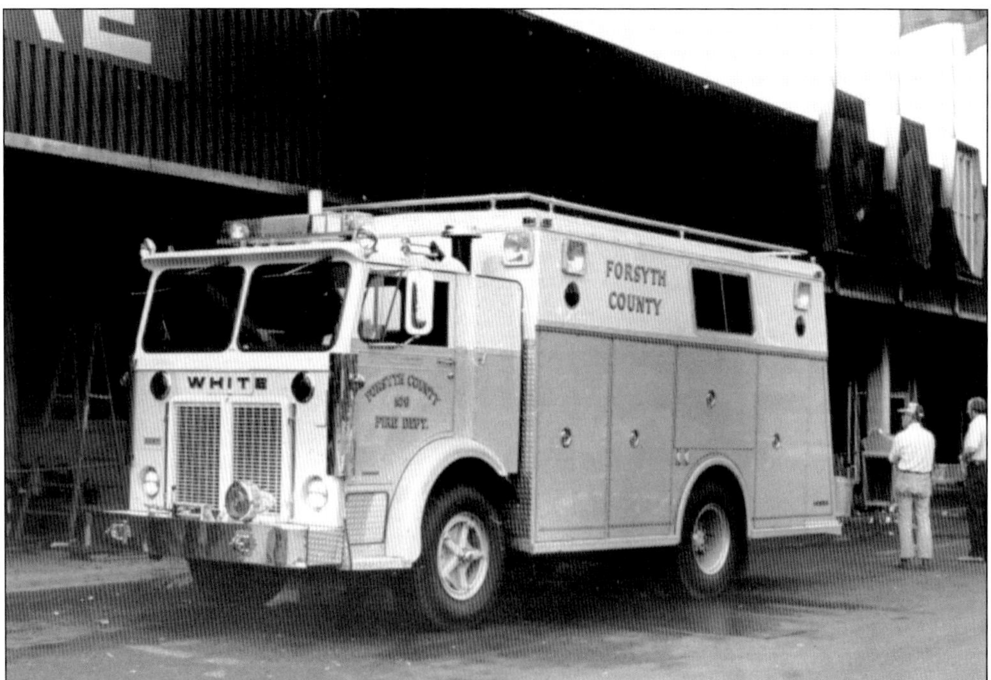

Forsyth County's 109 truck was now a 1975 White/Saulsbury heavy rescue truck. The Forsyth County 109 truck has always evolved to meet the needs of the volunteer and combination fire departments. (Courtesy of FCFHS.)

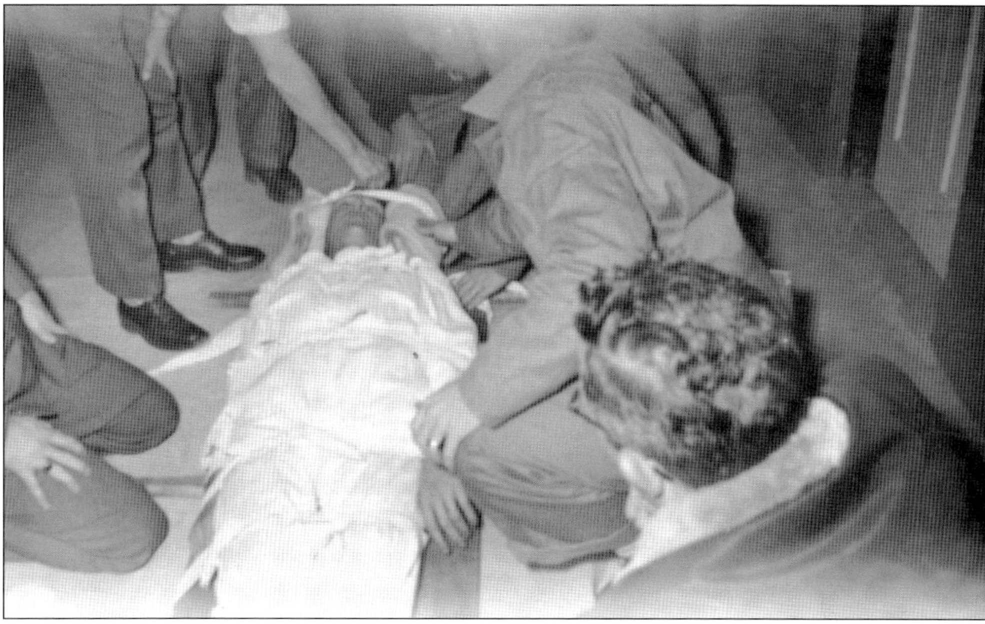

Pictured here are volunteer fire department members training on patient packaging for medical calls in 1968. The volunteer firefighter was now not only an expert in firefighting, but also took on the role of performing emergency medical care. (Courtesy of FCFHS.)

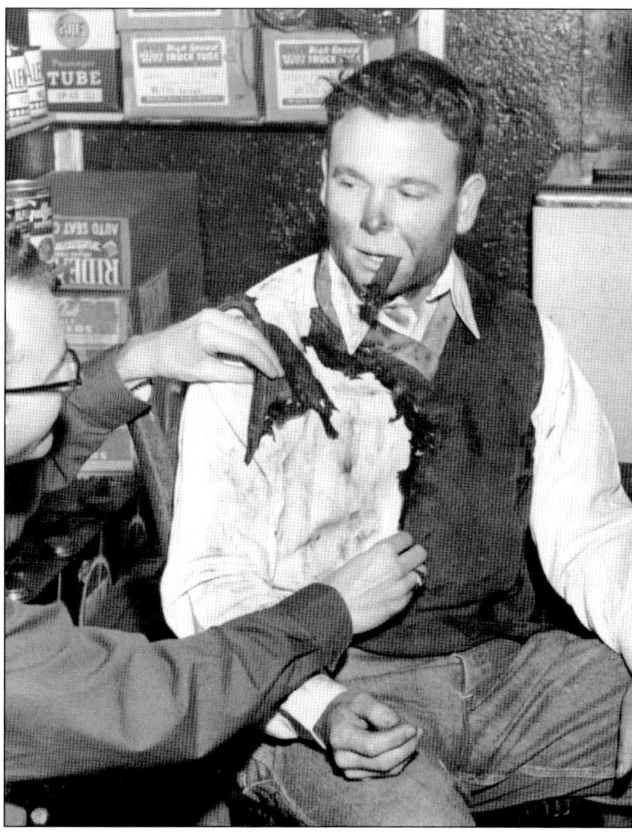

A Walkertown Volunteer Fire Department member is performing a patient assessment on Rick Waggoner after an accident at a service station. (Courtesy of FCFHS.)

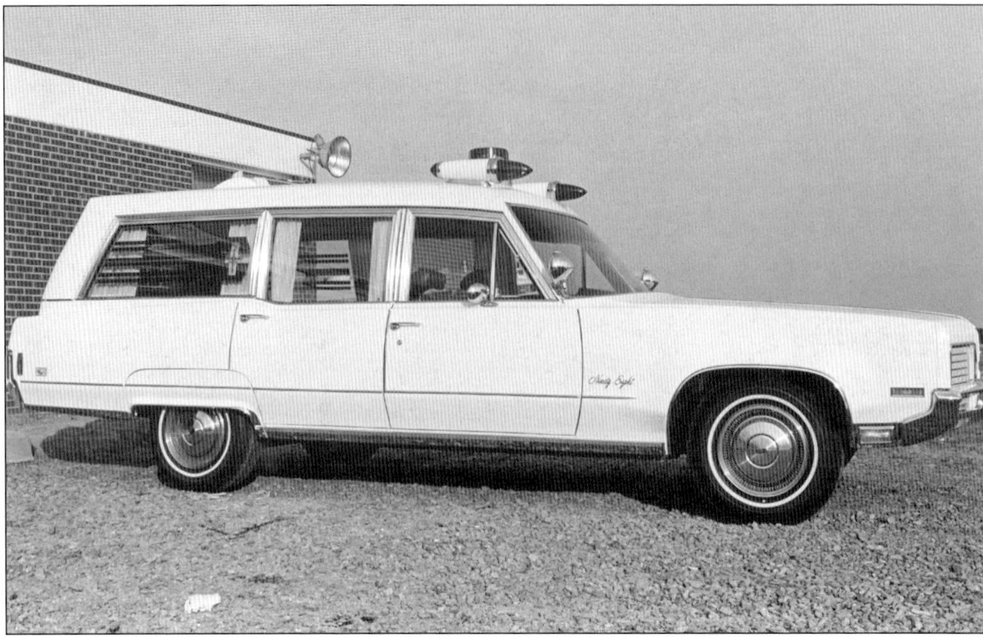

Before emergency medical services, local funeral homes provided the majority of ambulance service throughout Forsyth County. Pictured is one of Forsyth County Ambulance Services' ambulances—a 1971 Oldsmobile 88. (Courtesy of FCFHS.)

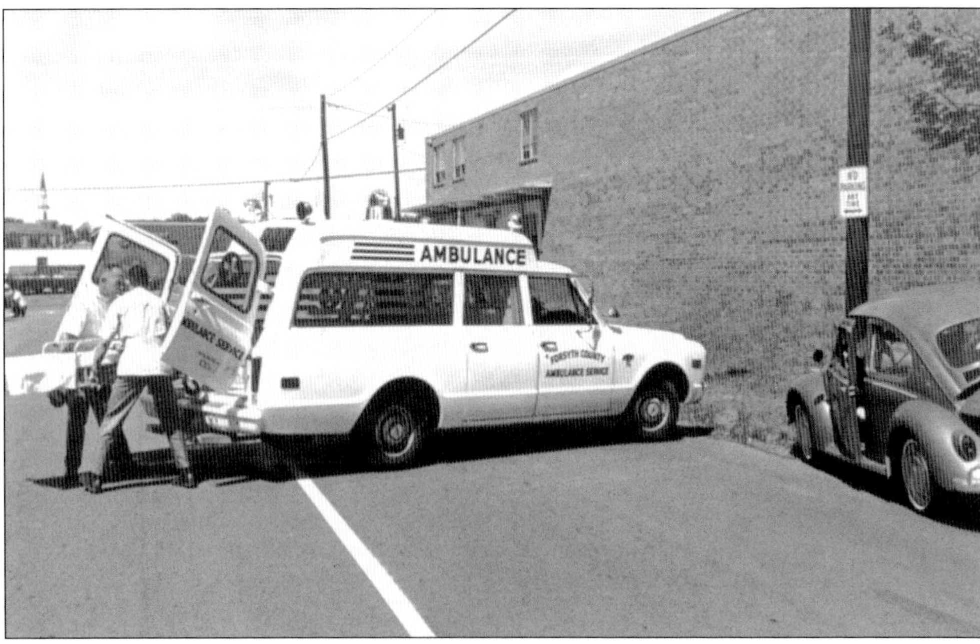

The Forsyth County Ambulance Service was established by the Forsyth County Board of Commissioners in January 1968. Two local funeral homes donated used ambulances and provided support. The Forsyth County Ambulance Service started with five vehicles operated by trained attendants. It was based out of the Forsyth County Fire Department station at Smith Reynolds Airport. Their headquarters would later move to Reynolds Memorial Hospital in 1974. (Courtesy of FCFHS.)

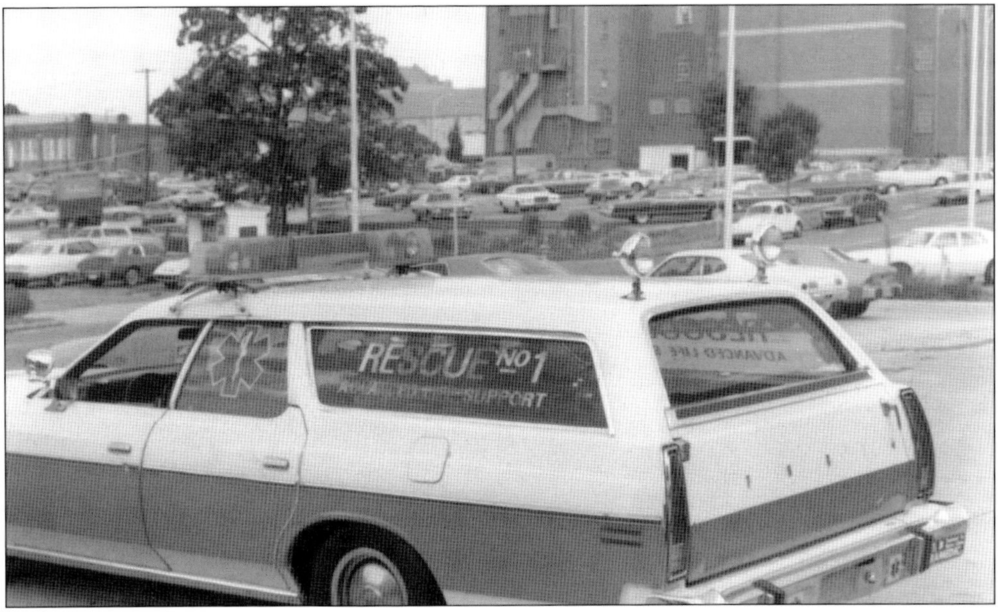

Forsyth County Emergency Medical Services is credited with graduating the first paramedic class in the state of North Carolina in 1975. This class was sponsored by the Forsyth Technical Institute, Bowman Gray School of Medicine, and Forsyth Memorial Hospital. The primary instructor of the class was Dr. L.W. Stringer. Pictured is a Forsyth County advanced life support (ALS) unit. ALS units like this were strategically placed throughout the county to provide medical services on an advanced prehospital level. (Courtesy of FCFHS.)

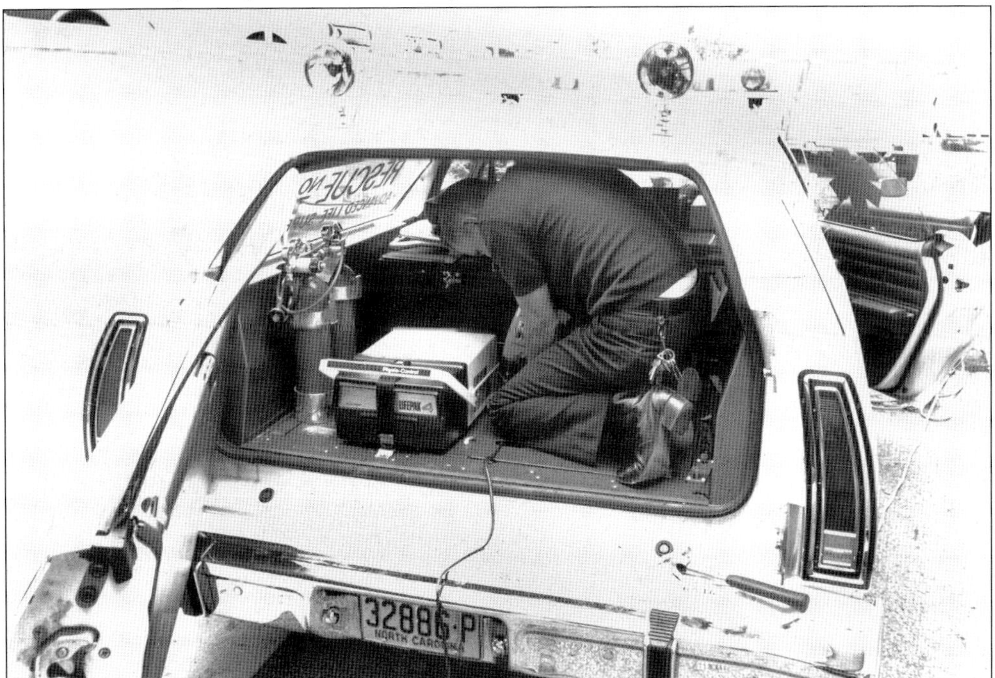

Paramedic Jack Hutchins is performing a truck check off in the back of an ALS support station wagon. (Courtesy of FCFHS.)

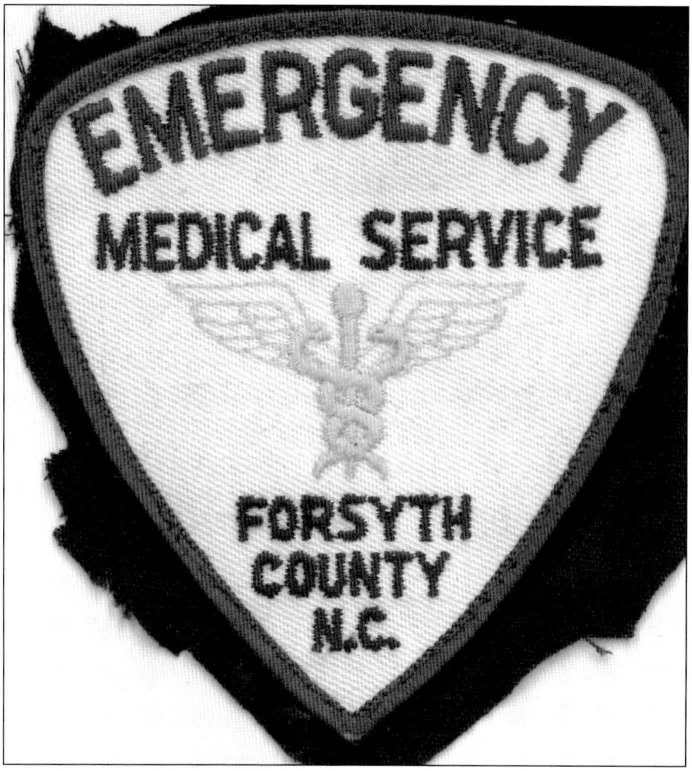

This is the Forsyth County Emergency Medical Services' first patch. (Courtesy of FCFHS.)

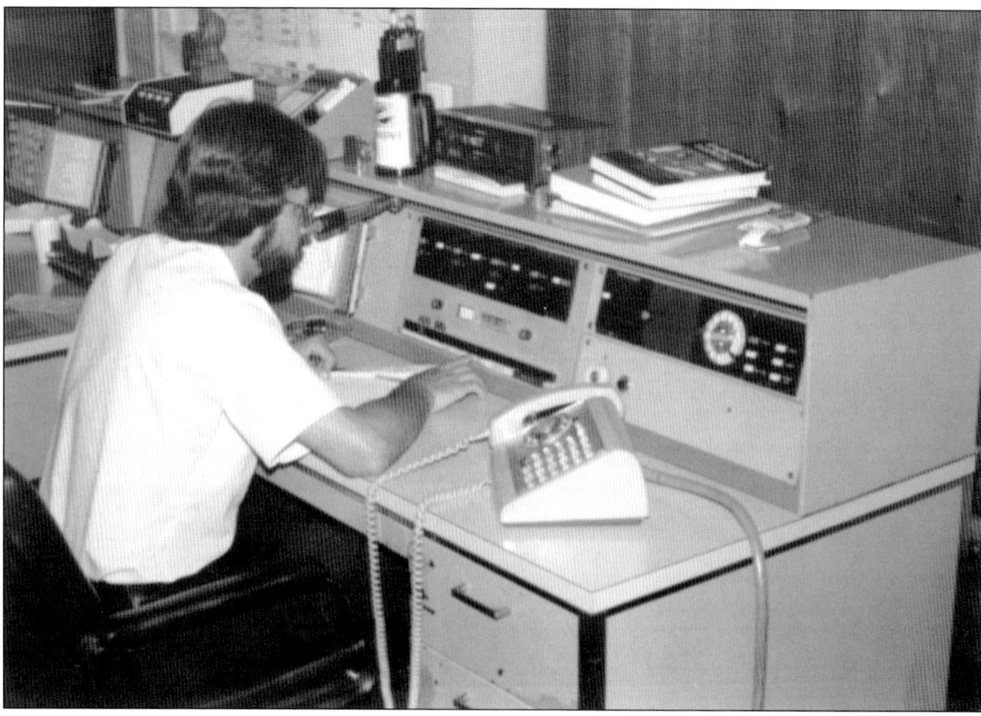

Pictured here is communications division radio operator Ray Deans working to dispatch ambulance units in Forsyth County. (Courtesy of FCFHS.)

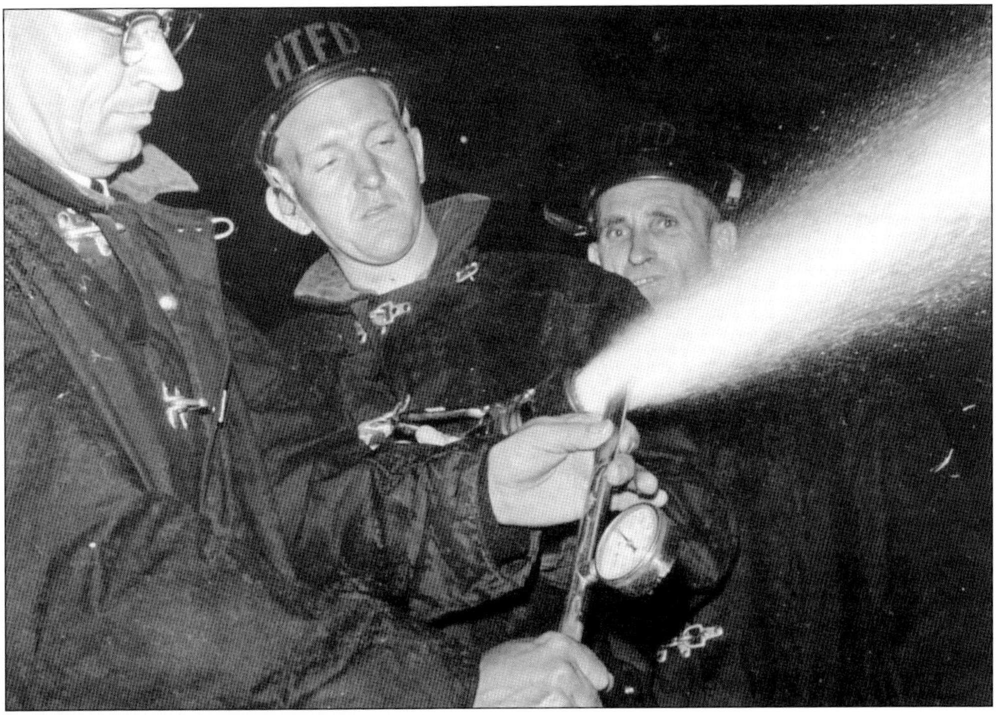

From left to right, Chief Roger Canada of the Horneytown Volunteer Fire Department is pictured with firefighters Bob Vaden and Vernon Shuskey demonstrating the use of a pitot gauge in measuring the nozzle pressure of a fire stream in 1958. (Courtesy of FCFHS.)

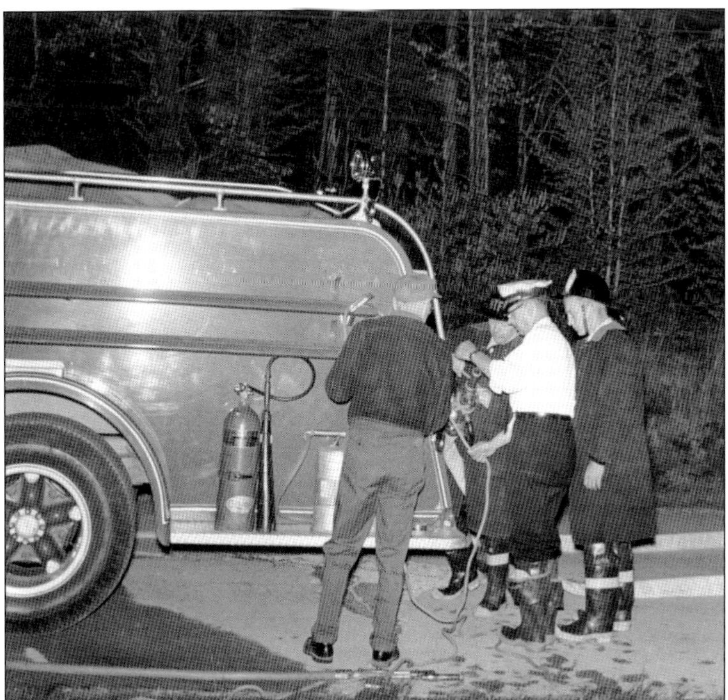

Pictured is Chief Roger Canada (in white) assisting Horneytown volunteer members (from left to right) Tom Andrews, Jim Venable, and Vernon Shuskey as they load hose on the department apparatus in 1958. (Courtesy of FCFHS.)

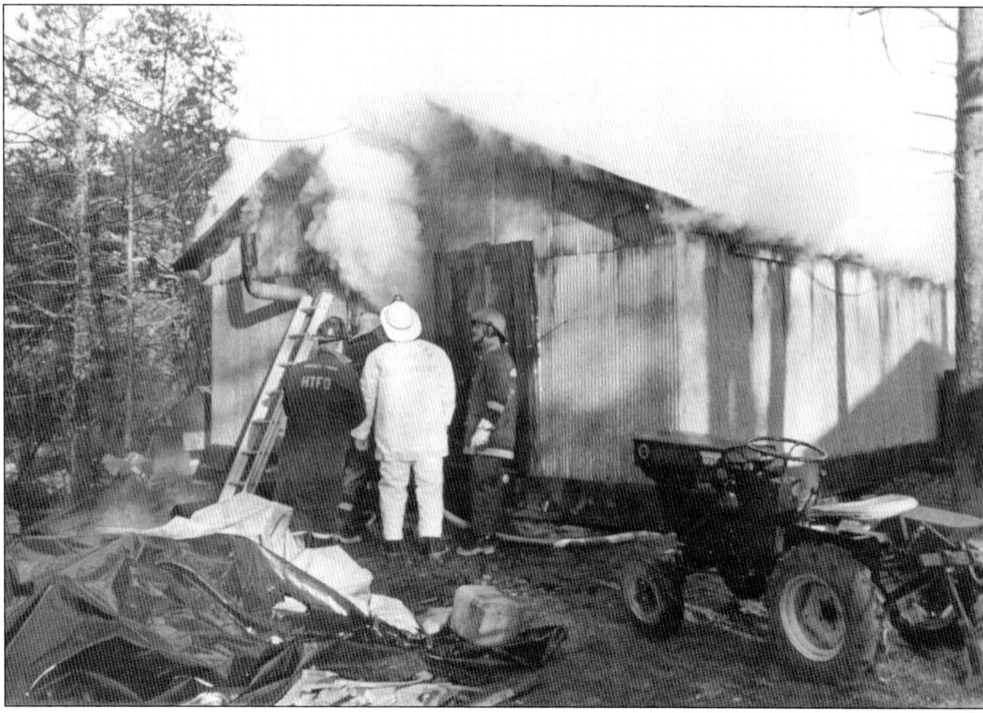

Forsyth County 109 personnel are shown assisting volunteer members on the scene of an outbuilding fire in the Horneytown fire district in 1975. (Courtesy of FCFHS.)

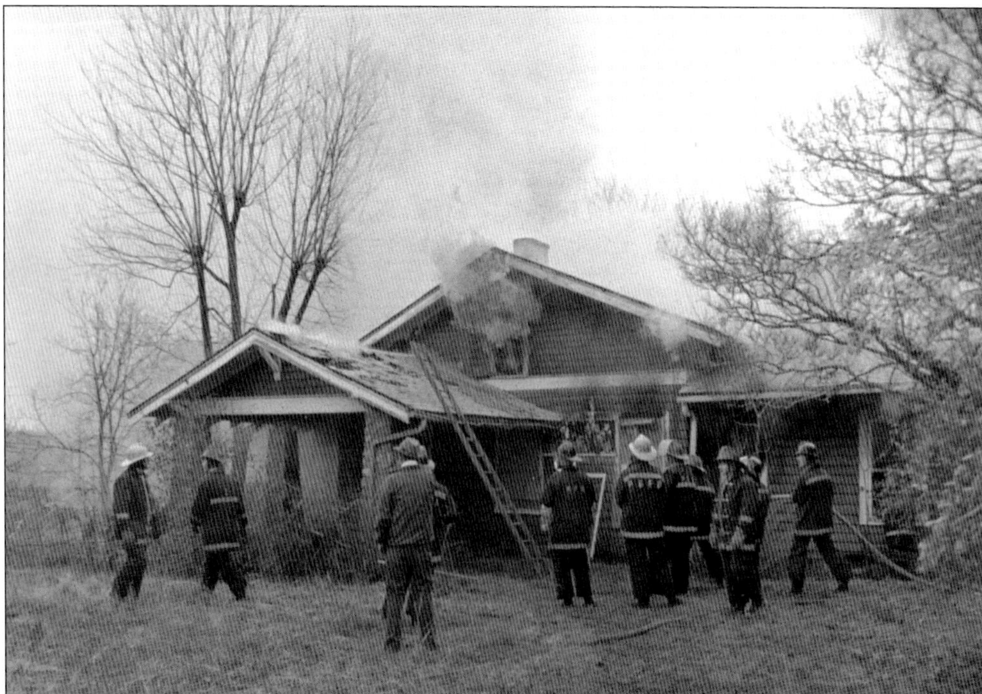

Piney Grove Volunteer Fire Department personnel are seen working on a training burn with the Forsyth County Fire Department in 1978. (Courtesy of FCFHS.)

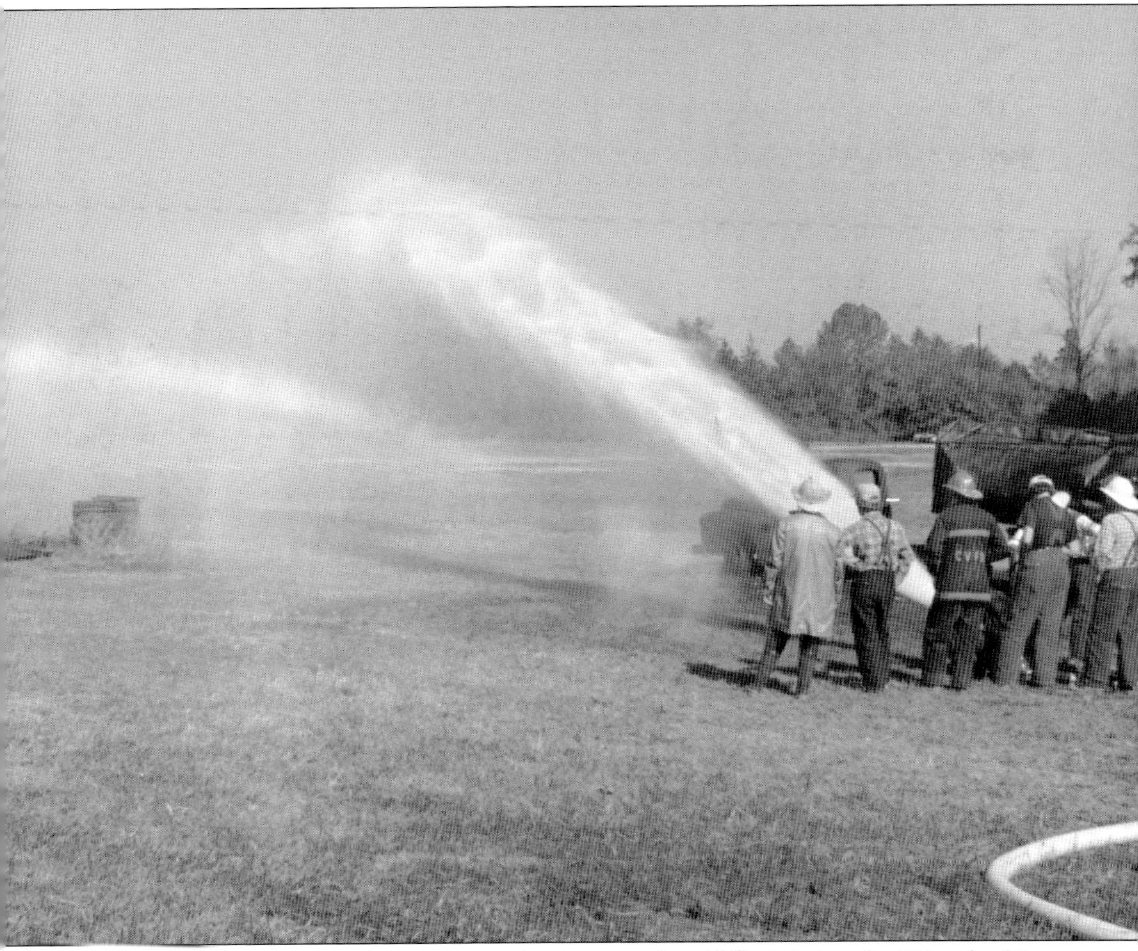

Forsyth County Fire Department personnel are pictured in 1961 assisting the Triangle, City View, Walkertown, and Talley's Crossing Volunteer Fire Departments as they tested a fire hose. (Courtesy of FCFHS.)

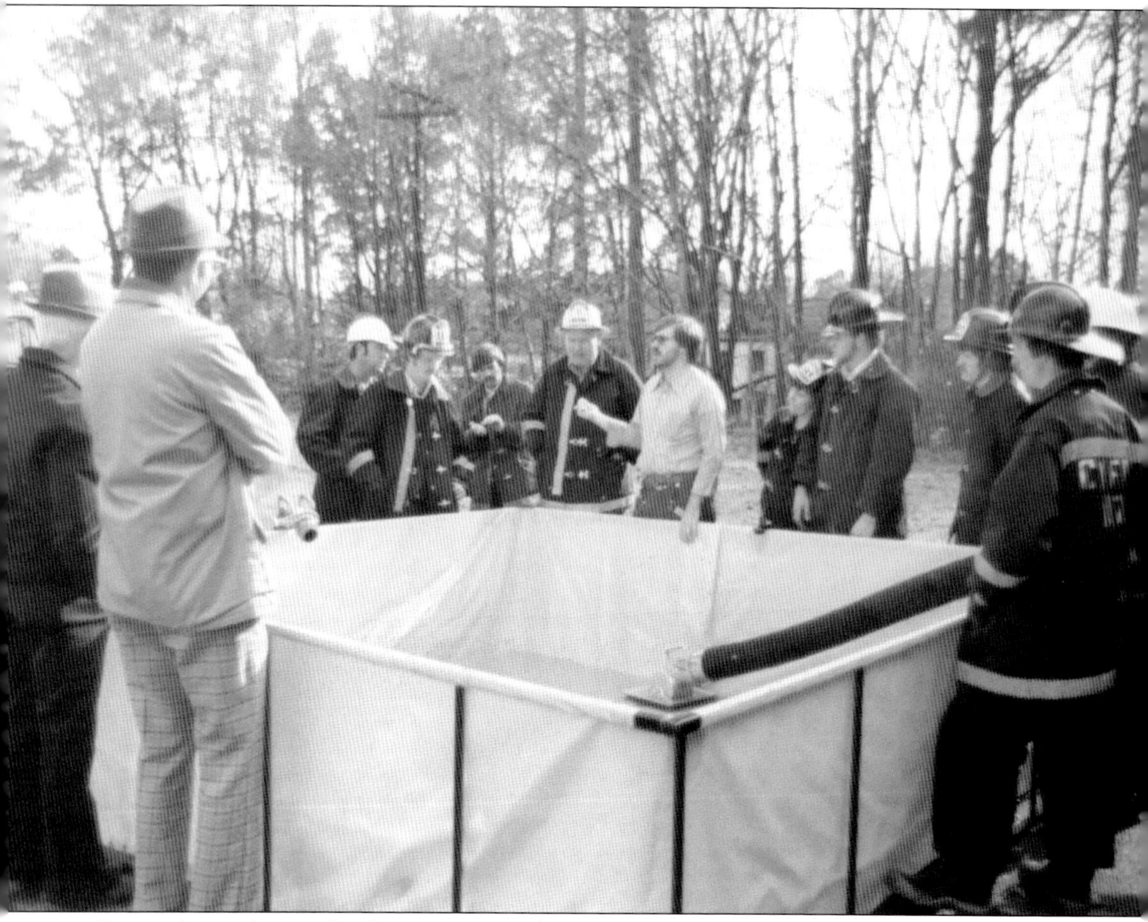

Pictured are volunteer fire department members training at Petree Elementary (formerly City View School) on drop tank operations. From left to right are two unidentified (backs to camera), Roger Wilmer, Doug Parrish, Franci Hubbard, Jerry Lewis, Emory Fulp, Robert Foster, David Vanhoy, Richard Coley Sr., and two unidentified. (Courtesy of FCFHS.)

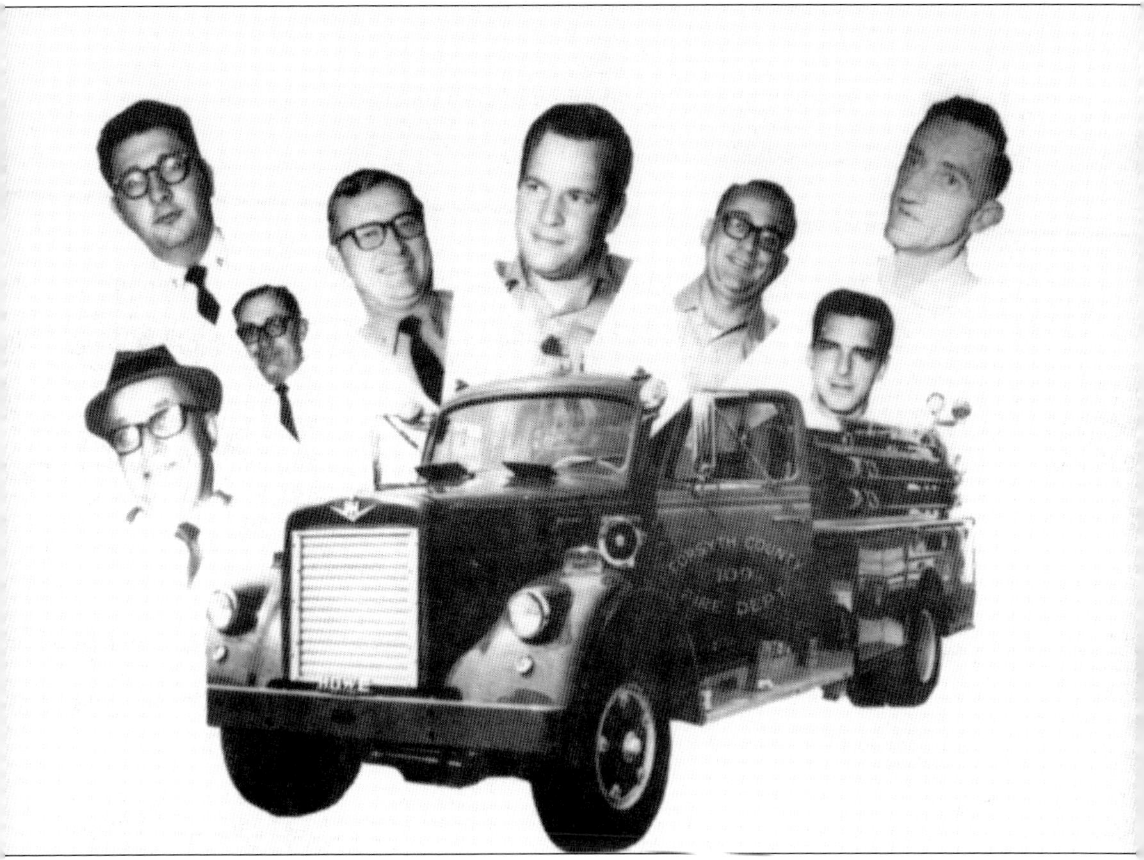

Pictured are several of the first pioneer firefighters for Forsyth County's 109 truck. From left to right are Reese Bauguess, Dick Hastings, Tom Andrews, Tink Shaffer, Mike Culbreth, John Conrad, John Hampton, and Tommy Johnson. These men always answered the call and adapted to and overcame whatever obstacle was presented to them. They were part of something much larger than themselves, and paved the way for future generations of firefighters, emergency operators, and emergency medical providers. (Courtesy of FCFHS.)

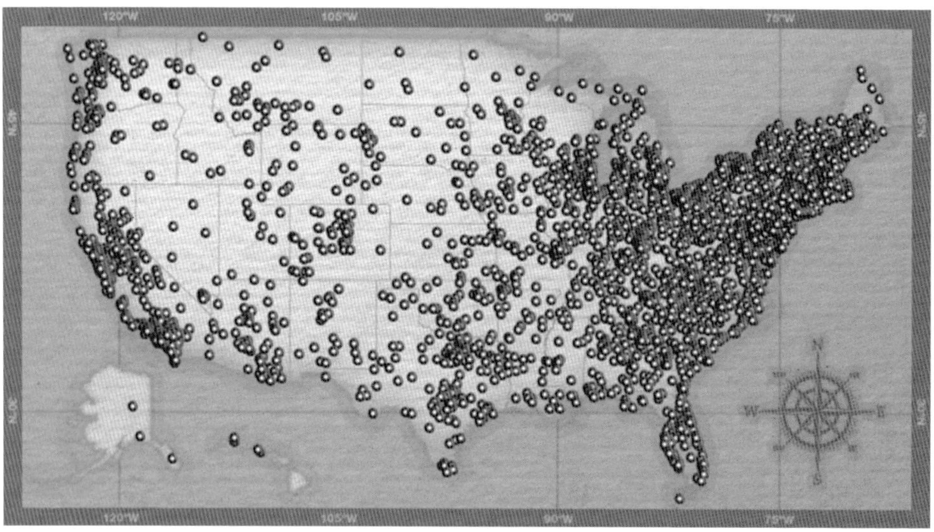